纺织服装教育"十四五"部委级规划教材

服装专业英语

（第三版）

GARMENT ENGLISH

辛芳芳 编著

東華大學出版社

内容提要

本书的编写是基于服装专业的大部分专业核心内容,涉及服装的起源、设计、结构、工艺、营销、服装展览、模特、流行趋势以及服装CAD等众多内容。

按照章节分类,本书共分为七章,每章内容由若干篇主文和泛读课文组成。面向大中专院校的服装专业英语的教学需要,本书编写中参阅了大量的最新专业英文资料,所节选的课文均经过仔细的比较和挑选,力争在有限的篇幅中,尽可能充分体现行业的主要内容和专业英语的规范表达方式;每篇核心课文后备有大量的专业词汇和重点语句的注释,帮助读者理解和学习。

本书内容具有较高的概括性和代表性,通过本书的学习,阅读者可以了解并掌握服装行业的结构和特点,既适合作为大中专院校专业学习教材,也可作为服装专业人员、业余爱好者的专业参考书。

本书编写中,一定存在某些不足之处,恳请各位行业专家、同仁以及读者在使用中提出宝贵意见,以便编者在以后的工作中改进。

编者　辛芳芳

图书在版编目(CIP)数据

服装专业英语/辛芳芳编著. —3版. —上海:东华大学出版社,2022.7
ISBN 978-7-5669-2078-2

Ⅰ.①服…　Ⅱ.①辛…　Ⅲ.①服装—英语—高等学校—教材
Ⅳ.①TS941

中国版本图书馆CIP数据核字(2022)第101483号

责任编辑　谢　未
封面设计　王　丽

服装专业英语(第三版)

辛芳芳　编著

东华大学出版社出版
上海市延安西路1882号
邮政编码:200051　电话:(021)62193056
上海普顺印刷包装有限公司
开本:787×1092　1/16　印张:13.5　字数:346千字
2022年7月第3版　2022年7月第1次印刷
ISBN 978-7-5669-2078-2
定价:45.00元

CONTENTS 目录

Chapter 1 THE INTRODUCTION OF CLOTHING / 服装概论

LESSON 1 CLOTHING / 服装

Clothing, coverings and garments intended to be worn on the human body. The words cloth and clothing are related, the first meaning fabric or textile, and the second meaning fabrics used to cover the body. The earliest garments were made of leather and other non-fabrics, rather than of cloth, but these non-fabric garments are included in the category of clothing①.

Fashion refers to the kinds of clothing that are in a desirable style at a particular time. At different times in history, fashionable dress has taken very different forms. In modern times nearly everyone follows fashion to some extent. A young woman would look odd if she wore the clothing that her grandmother had worn when young②. However, only a small minority of people dresses in the clothing that appears in high-fashion magazines or on fashion-show runways.

It is not always easy to tell the difference between basic clothing and fashionable clothing. Especially today, fashion designers often use inexpensive and functional items of clothing as inspiration. Blue jeans, for instance, originated as functional work clothing for miners and farmers. Yet today, even people who dress in jeans, T-shirts, and sports clothes may be influenced by fashion③. One year, fashionable jeans may have narrow legs; the next year the legs may be baggy.

Clothing historians trace the development of dress by studying various sources, including magazines and catalogs, paintings and photographs, and hats, shoes, and other surviving items. Reliable evidence about everyday clothing from the past can be hard to obtain because most publications and images concern the fashions of the wealthy④. Furthermore, clothing that has survived from the past tends not to be typical of what was worn in daily life. Museum collections are full of fashionable ball gowns, for example, but have very few everyday dresses worn by ordinary working-class women⑤. Even fewer examples of ordinary men's clothing have been saved. Images, such as paintings, prints, and photographs, do provide considerable evidence of the history of everyday clothing. These sources

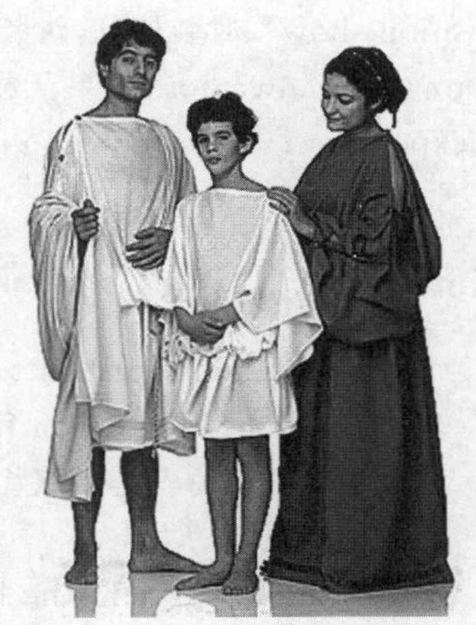

图 1 The ancient Greeks' apparel

indicate that although everyday clothing does not usually change as rapidly as fashionable dress, it does change constantly.

Words and Phrases

clothing [ˈkləuðiŋ] *n*. 服装，服饰，衣着，衣服，衣饰
garment [ˈgaːmənt] *n*. 衣服，服装
wear [wεə] *n*. 服装，衣服，穿戴物
cloth [klɔ(ː)θ] *n*. 织物，布，衣料
fabric [ˈfæbrik] *n*. 织物，织品，面料，布
textile [ˈtekstail] *n*. 纺织品，织物，纺织的，纺织原料
leather [ˈleðə] *n*. 皮革，皮革制品
category [ˈkætigəri] *n*. 种类，部属；类目
desirable [diˈzaiərəbl] *adj*. 理想的，令人满意的，良好的，优良的
style [stail] *n*. 型，款式，式样；时尚；类型；气派，风度，格调
fashionable [ˈfæʃənəbl] *n*. 时髦人物，流行的 *adj*. 时尚的
dress [dres] *v*. 穿着；*n*. 服装，礼服，连衣裙
form [fɔːm] *n*. 造型，外形，体型；人体模型
odd [ɔd] *adj*. 奇特的，古怪的
minority [maiˈnɔriti] *adj*. 少数，少数的
fashion-show 时装展览，时装表演
runway [ˈrʌnwei] *n*. 时装天桥
inexpensive [ˌiniksˈpensiv] *adj*. 廉价的，便宜的
functional [ˈfʌŋkʃənl] *adj*. 机能的，功能的
item [ˈaitem] *n*. 项；条款；项目；产品；展品
inspiration [ˌinspəˈreiʃən] *n*. 灵感
jeans [dʒiːnz] *n*. 牛仔裤，紧身裤，粗斜纹棉布裤
originate [əˈridʒineit] *v*. 产生，引起
T-shirt [ˈtiː ˌʃəːt] *n*. T 恤衫，短袖圆领汗衫
legs [legs] *n*. 裤脚
baggy [ˈbægi] *adj*. 膨胀的，凸出的
historian [hisˈtɔːriən] *n*. 历史学家
catalog [ˈkætəlɔg] *n*. 目录
hat [hæt] *n*. 帽子
shoe [ʃuː] *n*. 鞋子
collection [kəˈlekʃən] *n*. 服饰系列，季节服装系列；时装展览，时装发布会
ball gown 晚会礼服，正规礼服
working-class 工薪族
ordinary [ˈɔːdinəri] *adj*. 原始的，普通的，平凡的
image [ˈimidʒ] *n*. 影像，肖像，图像，形象，反映
constantly [ˈkɔnstəntli] *adv*. 不变地，不断地，时常地

Notes

① The earliest garments were made of leather and other non-fabrics, rather than of cloth, but these non-fabric garments are included in the category of clothing.
最早的衣物不是用织物，而是用皮革和其他非织物制作的，但是这些非织造衣物也属于服装的大类。

② Fashion refers to the kinds of clothing that are in a desirable style at a particular time. At different times in history, fashionable dress has taken very different forms. In modern times nearly everyone follows fashion to some extent. A young woman would look odd if she wore the clothing that her grandmother had worn when young.

时装是指特定时期在款式上受欢迎的各种服装。在不同的历史阶段，流行服装的造型差别很大。在现代，几乎每个人都在某种程度上追逐时尚。如果年轻女士穿上她祖母年轻时所穿的衣服，就会显得怪异。

③ Especially today, fashion designers often use inexpensive and functional items of clothing as inspiration. Blue jeans, for instance, originated as functional work clothing for miners and farmers. Yet today, even people who dress in jeans, T-shirts, and sports clothes may be influenced by fashion.

尤其是今天，时装设计师的设计灵感常来自于廉价面料和服装的功能性。例如，蓝色牛仔装早期是矿工和农夫的工作服。然而今天，即便是 T 恤衫、牛仔裤和运动装的简单搭配也受到时尚的影响。

④ Clothing historians trace the development of dress by studying various sources, including magazines and catalogs, paintings and photographs, and hats, shoes, and other surviving items. Reliable evidence about everyday clothing from the past can be hard to obtain because most publications and images concern the fashions of the wealthy.

服装史学者通过多种方式来研究衣着的发展，如杂志和分类目录、绘画和照片、帽、鞋以及其他留存下来的东西。有关以往日常衣服的可靠证据很难获得，因为大多数的出版物和图片资料都只关注富人们的时装。

⑤ Furthermore, clothing that has survived from the past tends not to be typical of what was worn in daily life. Museum collections are full of fashionable ball gowns, for example, but have very few everyday dresses worn by ordinary working-class women.

此外，过去留存下来的服装也并非是典型的日常装，例如博物馆里藏有大量的高级晚礼服，而普通工薪阶层妇女的日常装却非常少。

Discussion Questions

1. List the reason why people wear clothes.
2. Explain how clothes reflect the way that people think and live in a society.

EXTENSIVE READING

ORIGIN AND HISTORY OF CLOTHING

According to archaeologists and anthropologists, the earliest clothing probably consisted of fur, leather, leaves or grass, draped, wrapped or tied about the body for protection from the elements. Knowledge of such clothing remains inferential, since clothing materials deteriorate quickly compared to stone, bone, shell and metal artifacts. Archeologists have identified very early sewing needles of bone and ivory from about 30,000 BC, found near Kostenki, Russia, in 1988.

Some human cultures, such as the various peoples of the Arctic Circle, until recently made their clothing entirely of furs and skins, cutting clothing to fit and decorating lavishly.

Other cultures have supplemented or replaced leather and skins with cloth: woven, knitted, or twined from various animal and vegetable fibers.

图 2 Sculptures of ancient costumes

Although modern consumers take clothing for granted, making the fabrics that go into clothing is not easy. One sign of this is that the textile industry was the first to be mechanized during the Industrial Revolution①; before the invention of the power-loom, textile production was a tedious and labor-intensive process. Therefore, methods were developed for making most efficient use of textiles.

One approach simply involves draping the cloth. Many people wore, and still wear, garments consisting of rectangles of cloth wrapped to fit, for example, the Scottish kilt or the Javanese sarong. Pins or belts hold the garments in place. The precious cloth remains uncut, and people of various sizes can wear the garment.

Another approach involves cutting and sewing the cloth, but using every bit of the cloth rectangle in constructing the clothing. The tailor may cut triangular pieces from one corner of the cloth, and then add them elsewhere as gussets. Traditional European patterns for men's shirts and women's chemises take this approach.

Modern European fashion treats cloth much more prodigally, typically cutting in such a way as to leave various odd-shaped cloth remnants. Industrial sewing operations sell these as waste; home sewers may turn them into quilts.

图 3 The original human activities

In the thousands of years that humans have spent constructing clothing, they have created an astonishing array of styles, many of which we can reconstruct from surviving garments, photos, paintings, mosaics, etc., as well as from written descriptions. Costume history serves as a source of inspiration to current fashion designers, as well as a topic of professional interest to costumers constructing for plays, films, television, and historical reenactment.

Words and Phrases

archaeologist [aːkiə'lɔdʒist] *n.* 考古学家

anthropologist [ænθrə'pɔlədʒist] *n.* 人类学家

fur [fəː] *n.* 毛皮

leather ['leðə] *n.* 皮革,皮革制品

drape [dreip] *v.* 垂坠,悬垂,立体裁剪

wrap [ræp] *v.* 包裹,围裹

tie [tai] *v.* 系,打结,扎,绑,捆

inferential [ˌinfəˈrenʃəl] *adj*. 推理的，可以推论的

deteriorate [diˈtiəriəreit] *v*. 损坏，损耗，变质

compare to　与……相比

shell [ʃel] *n*. 贝壳，壳

artifact [ˈaːtifækt] *n*. 文化遗物，遗迹

identify [aiˈdentifai] *v*. 认出；识别；鉴别；验明

ivory [ˈaivəri] *n*. 象牙；(海象等的)长牙

the Arctic Circle　北极圈

skin [skin] *n*. 毛皮，兽皮

fit [fit] *v*. 合身，合体，使合身

decorate [ˈdekəreit] *v*. 装饰

lavishly [ˈlæviʃli] *adv*. 丰富地，浪费地

supplement [ˈsʌplimənt] *n*. 增补(物)，补充(物)

weave [wiːv] *n*. 梭织，织物，织法，编织式样　*v*. 编织

knit [nit] *n*. 针织，针织品，针织服装　*v*. 针织

twine [twain] *v*. 合股，搓，交织，缠绕

fiber [ˈfaibə] *n*. 纤维

take for granted　认为……理所当然

mechanize [ˈmekənaiz] *v*. 机械化，机械化(生产)

power loom　动力织布机

tedious [ˈtiːdiəs] *adj*. 单调乏味的，令人生厌的，繁重的

labor-intensive　劳动密集型的

Scottish kilt　苏格兰褶裥短裙

Javanese sarong　爪哇莎笼围裙

pin [pin] *n*. 大头针，别针

belt [belt] *n*. 带，皮带，腰带，肩带

uncut [ˈʌnˈkʌt] *adj*. 不裁剪的，未经裁剪的

rectangle [ˈrektæŋgl] *n*. 矩形，长方形

triangular [traiˈæŋgjulə] *adj*. 三角形的

gusset [ˈgʌsit] *n*. 三角形衬料

chemise [ʃiˈmiːz] *n*. 直统连衣裙，女式无袖衬衣

prodigally [ˈprədigli] *adv*. 浪费地

odd-shaped　不规则形状的

remnant [ˈremnənt] *n*. 剩余，零料

waste [weist] *n*. 废料，废纱

quilt [kwilt] *n*. 被子　*v*. 绗缝

astonish [əsˈtəniʃ] *v*. 使……吃惊

an array of　一批，一系列

reconstruct [ˈriːkənˈstrʌkt] *v*. 重现，再现

mosaic [məˈzeiik] *n*. 马赛克，马赛克(图案)

description [disˈkripʃən] *n*. 叙述，描写，描绘

reenactment [riːiˈnæktmənt] *n*. 重演

Notes

① Industrial Revolution widespread replacement of manual labor by a machine that began in Britain in the 18th century and is still continuing in some parts of the world.
始于 18 世纪，起源自英国的工业革命，运用机器大生产替代了手工生产。在世界某些范围内，这种替代一直在持续。

LESSON 2　FUNCTIONS OF CLOTHING / 服装的功能

Since prehistoric times, people in almost all societies have worn some kind of clothing. Many theories have been advanced as to why humans began to wear clothing. Some argued that the origin of clothing was functional — to protect the body from the environment. Others argued that some clothing was designed for sexual attraction — to display the body's beauty.

图 4 Eskimo wear seal clothing and rein deer pelts from cold weather

Today, modern scholars believe that except the two functions clothing still provides a mark of identity and a means of nonverbal communication. In traditional societies, clothing functions almost as a language that can indicate a person's age, gender, marital status, place of origin, religion, social status, or occupation. In modern industrialized societies, clothing is not so rigidly regulated and people have more freedom to choose which messages they wish to convey. Nevertheless, clothing can still provide considerable information about the wearer, including individual personality, economic standing, even the nature of events attended by the wearer①.

A society's economic structure and its culture, or traditions and way of life, also influence the clothing that its people wear. In many societies, religious laws regulated personal behavior and permitted only members of an elite class to wear certain prestigious items of clothing. Even in modern democracies, clothing may represent social standing. Clothing with a designer label tends to be relatively expensive, so it may function as an outward sign of a person's economic standing②. Clothing most obviously defines a social role in the case of uniforms, such as those worn by police officers and nurses, and garments worn by clergy or members of religious orders.

图 5 Arabian robe hide the body from the hot sunlight

图 6 The miner's clothing with some functions

Clothing also derives meaning from the environment in which it is worn. In most cultures brides and grooms as well as wedding guests wear special clothes to celebrate the occasion of a marriage. The clothing worn for rituals such as weddings, graduations, and funerals tends to be formal and governed by unwritten rules that members of the society agree upon③. Clothing may also signal participation in leisure activities. Certain types of recreation, especially active sports, may require specialized clothing. For example, football, soccer, and hockey players wear matching jerseys and pants designed to accommodate such accessories as protective pads.

Most modern societies comprise different social groups, and each group has its own beliefs and behaviors. As a result, different clothing subcultures exist.

Words and Phrases

prehistoric [ˈpriːhisˈtɔrik] *adj*. 史前的
nonverbal [ˈnɔnˈvəːbəl] *adj*. 非语言的
gender [ˈdʒendə] *n*. 性别
marital status 婚姻状态
industrialize [inˈdʌstriəlaiz] *v*. 工业化
rigidly [ˈridʒidli] *adj*. 严格的
regulate [ˈregjuleit] *v*. 管理，控制
considerable [kənˈsidərəbl] *adj*. 重要的，不可忽视的
individual [ˌindiˈvidjuəl] *adj*. 个人的，个别的，单独的，个性的
standing [ˈstændiŋ] *n*. 地位，身份，名声
elite [iˈliːt] *n*. 精英，中坚
prestigious [presˈtiːdʒəs] *adj*. 有威望，有声誉的
democracy [diˈmɔkrəsi] *n*. 民主
represent [riːpriˈzent] *v*. 表现，表示，代表
outward [ˈautwəd] *adj*. 外面的，明显的，公开的
clergy [ˈkləːdʒi] *n*. 牧师，僧侣，神职人员
derive from 由来，起源自……
bride [braid] *n*. 新娘
groom [grum] *n*. 新郎
ritual [ˈritjuəl] *n*. 仪式，典礼
graduation [grædjuˈeiʃən] *n*. 毕业典礼
funeral [ˈfjuːnərəl] *n*. 葬礼
participation [paːˌtisiˈpeiʃən] *v*. 参加，参与
recreation [rekriˈeiʃ(ə)n] *n*. 娱乐，消遣
hockey [ˈhɔki] *n*. 曲棍球，冰球
matching [ˈmætʃiŋ] *adj*. 相配的
jersey [ˈdʒɜːzi] *n*. 紧身运动套衫；平针织物
accommodate [əˈkɔmədeit] *v*. 使适应，向……提供
accessory [ækˈsesəri] *n*. 服饰品；配件，备件；辅助设备
pad [pæd] *n*. 衬垫，肩垫
comprise [kəmˈpraiz] *v*. 包含，由……组成
subculture [ˈsʌbˌkʌltʃə] *n*. 亚文化

Notes

① In traditional societies, clothing functions almost as a language that can indicate a person's age, gender, marital status, place of origin, religion, social status, or occupation. In modern industrialized societies, clothing is not so rigidly regulated and people have more freedom to choose which messages they wish to convey. Nevertheless, clothing can still provide considerable information about the wearer, including individual personality, economic standing, even the nature of events attended by the wearer.

在传统社会，服装几乎可以作为一种语言，以表明一个人的年龄、性别、婚姻状况、出生地、信仰、社会地位和职业。在现代工业化社会，服装不再受如此严格的限制，人们在选择所希望传递的信息方面有了更大的自由。即便如此，服装仍能展示穿着者的众多信息，诸如其个性、经济地位，甚至是穿着者所出席场合的性质等。

② In many societies, religious laws regulated personal behavior and permitted only members of an elite class to wear certain prestigious items of clothing. Even in modern democracies, clothing may represent social standing. Clothing with a designer label tends to be relatively expensive, so it may function as an outward sign of a person's economic standing.

在许多社会,教规规范教徒们的行为,并规定精英阶层才能穿着某些特权服饰。即使在现代民主社会,服饰也代表着社会地位,贴有设计师标签的服饰价格相对昂贵,因此它具有了表现经济地位的功能。

③ Clothing also derives meaning from the environment in which it is worn. In most cultures brides and grooms as well as wedding guests wear special clothes to celebrate the occasion of a marriage. The clothing worn for rituals such as weddings, graduations, and funerals tends to be formal and governed by unwritten rules that members of the society agree upon.

服装也从穿着时所处的场合衍生出含义。在许多文化中,新娘、新郎以及婚礼来宾穿着专门的服装来庆祝结婚典礼。在婚礼、毕业典礼以及葬礼等仪式上,按社会成员所认同的不成文规矩,穿着的服装一般是正式的。

Discussion Questions

1. Give examples of what clothes can tell you about a person's occupation, nationality, and cultural heritage.
2. Discuss the influences that other cultures have on the clothes we wear today.

EXTENSIVE READING

THE FIRST ERA OF MODERN FASHION

During the first hundred years of modern fashion (from the 1860s to the 1960s), Paris was known as the center of innovation and set the annual trends followed by the rest of the world. Organized fashion shows on fixed dates began after World War I, an innovation that coincided with France's need for fashion as an export and the influx of professional buyers from the United States and other countries in Europe. The professional buyers, through a fee arrangement with the designers, acquired models for manufacturing at lower prices in their own countries.

The absolute dictatorship of fashion by Paris was undermined in the 1920s. Early in the decade Chanel① popularized "the Poor Look" of simple dresses, jersey suits, sweaters, cloche hats, and pants. Patou② introduced the sportswear approach to fashion which he described as follows: "I have aimed at making pleasant to the eye and allowing absolute liberty of movement". These look replaced the elaborate fashions and constricting stays that kept women sedentary with a new aesthetic ideal for the modern woman-slim, active, athletic. Chanel's "Poor Look" and Patou's sportswear were also much easier to imitate, thereby opening up fashionability to more consumers.

Daytime dress became more comfortable and functional, but evening fashion continued to be the epitome of seductive femininity. This fracturing of looks played out in ever more varied forms. A woman could choose to be a sexy woman, a "schoolgirl" in black dress with white collar and cuffs, a professional woman in a tailored suit, or a sporty woman in trousers and a sweater set. After the 1920s, the unity of a single fashion message disappeared and disparate and sometimes antagonistic looks shared the stage. Fashion gained transformative power as it became possible to manipulate appearance to express self, personality, and individuality-to change the way a woman saw herself and how other people saw her. Instead of issuing strict injunctions, fashion began to offer a diversified set of options inviting the consumer to choose.

Words and Phrases

trend [trend] *n.* 时尚，流行，趋势
innovation [inəu'veiʃən] *n.* 创新，革新，改革
coincide with　相符，与……一致
influx ['inflʌks] *v.* 汇集；涌进，涌入
professional buyer　专业采购员，专业买手
arrangement [ə'reindʒmənt] *n.* 商定，协议
dictatorship [dik'teitəʃip] *n.* 主宰，独裁
undermine [ʌndə'main] *v.* 在……下方
popularize ['pɔpjuləraiz] *v.* 推广
jersey suit　针织套装
sweater ['swetə] *n.* 运动衫；针织套衫；毛衣，毛线衫
cloche hat　钟形女帽
pants [pænts] *n.* 裤子，长裤，便裤
sportswear ['spɔːtswɛə] *n.* 便装；运动服装
elaborate [i'læbərət] *adj.* 精致的，精巧的
aesthetic [iːs'θetik] *adj.* 美学的，审美的，有美感的
sedentary ['sedəntəri] *adj.* 少动的；固定于一点的
constrict [kən'strikt] *v.* 压缩，收缩
stay [stei] *n.* 滚边，窄带
slim [slim] *adj.* 细长的，苗条的，纤细的
active ['æktiv] *adj.* 活跃的，积极的，精力充沛的
athletic [æθ'letik] *adj.* 运动的，身体健壮的，活跃的
imitate ['imiteit] *v.* 模仿；模拟；仿效；效法，仿造
fashionability ['fæʃənəbliti] *n.* 时尚，流行
epitome [i'pitəmi] *n.* 概要，缩影，象征
seductive [si'dʌktiv] *adj.* 诱惑的，引人注意的，有魅力的
femininity [femi'niniti] *n.* 女性气质，女人味
play out　放出，用完，结束
collar ['kɔlə] *n.* 领，衣领，上领
cuff ['kʌf] *n.* 袖口
raw material　原材料
man-made　人造的，合成的
tortoiseshell ['tɔːtəʃel] *n.* 龟甲，玳瑁
distribution [distri'bjuːʃən] *n.* 分发，分配
tailored suit　西式套装，精做西装
trousers ['traʊzəz] *n.* 裤子，长裤，西装裤
disparate ['dispərit] *adj.* 完全不同的，全异的
antagonistic [ænˌtægə'nistik] *adj.* 对抗的，不相容的
look [luk] *n.* 风貌，风格、型；款式，外表，姿态
transformative [træns'fɔːmətiv] *adj.* 使变化的，有变形力的
manipulate [mə'nipjuleit] *v.* 操纵，操作
self [self] *n.* 自己，自我；本性
personality [pəːsə'næliti] *n.* 人的存在；个

性，人格

individuality [indiˌvidjuˈæliti] *n*. 个体，个性

diversify [daiˈvəːsifai] *adj*. 形形色色的，多种多样的

issue [ˈisjuː] *v*. 颁布，发布，发行

injunction [inˈdʒʌŋkʃən] *n*. 命令，禁令

form [fɔːm] *n*. 造型，外形，体型；人体模型

Poor Look 破旧型款式，贫穷装

Notes

图 7 Coco Chanel

① Coco Chanel (1883～1971), French fashion designer and one of the leaders of haute couture (high fashion), whose name was synonymous with elegance and chic.
可可·夏奈尔(1883～1971)，法国时装设计师，优雅和别致的代名词，法国高级时装屋领袖之一。

② Jean Patou (1880～1936), French clothes designer, who opened a fashion house in 1919 and was an overnight success.
让·巴杜(1880～1936)，法国服装设计师，在 1919 年开设了自己的时装工作室，一夜成名。

Chapter 2 MATERIALS FOR CLOTHING / 服装材料

LESSON 3 FIBERS / 纺织纤维

Natural Fibers

Natural fibers come from animal or vegetable sources. All the natural fibers, except cultivated silk, have relatively short fibers which are combed and twisted to form yarn that is strong enough for use in the manufacture of the fabric① . Cultivated silk which is unwound from the silk moth's cocoon can be 2000 meters long and is therefore considered as continuous or a filament fiber.

Although frequently blended or woven together, many garments are made entirely from silk, wool, cotton and flax; the other fibers listed are usually mixed with the main fibers to add practical characteristics or aesthetic interest to the fabric② . When characteristics are added or suppressed by chemical processes and breeding, the structure of the fibers are not changed. Advances in present gene research are beginning to alter this position. Fabrics made from natural fibers, especially cotton, still hold a strong position in the market, despite the fact that they can be more expensive than a product made from man-made fibers. They are comfortable to wear because of their natural absorbency, and there is great aesthetic appeal in their textures, their dye affinities and their handle③.

Man-made Fibers

Man-made fibers are produced from chemical solutions that are manufactured into fibers; for example, a chemical liquid can be forced through minute holes and then solidified in air or by chemical processes④ . They can be used in filament or cut to form staple fibers. A fiber can be produced from a solution (regenerated fiber) that has a natural source or from a solely chemical or mineral source (synthetic fiber).

Regenerated cellulosic fibers are reconstituted by converting natural products such as wood pulp and cotton by solvents into a liquid form for spinning. Synthetic fibers are made from chemical sources. They are mainly petroleum based.

Man-made fibers began by copying the characteristics of natural fibers. Originally, man-made fiber lengths were matched to those of existing natural fibers because natural fibers were successful and the new fibers could be processed on existing machinery⑤. These regenerated cellulosic fibers, a chemical reduction of a natural source (wood pulp) created the first man-made fiber (rayon viscose) known as "artificial silk". Acetate followed, and more recently the new fiber lyocell has been created. The manufacture of synthetic fibers for the garment industry has now overtaken the production of all natural fibers. Nylon, polyester and acrylic originally displayed unique characteristics that were easy to identify. Until quite recently it was fairly easy to place the fibers of a fabric within a generic group and make certain assumptions about their properties; now, recognition is more difficult⑥.

The appearance, handle and comfort of a fabric are affected by the structure of the fibers. Whilst the length and external surface of the fiber is important, the internal structure also determines the basic properties of a particular fiber. The shape of the fiber can determine the luster: for example, the filaments of silk are prism shaped and reflect light. The cross-sections of fibers can be changed by varying the holes on the spinneret to match the shape of natural fibers or experiment with new shapes⑦. These can be round, cross-like, triangular, Y-shaped or bean-shaped. The structural shapes of fibers also determine more mechanical properties such as bulk, stiffness and absorbency; for example, circular shaped fibers tend to resist bending, Y-shaped fibers give resilience, hollow fibers are light in relation to their bulk. However, the yarn construction, fabric structure and finish have to be combined intelligently to satisfy aesthetic and practical market demands.

Words and Phrases

comb [kəum] *n.* 梳 *v.* 梳理
twist [twist] *v.* 捻,拧,编织
yarn [jaːn] *n.* 纱,纱线
cultivated silk 家蚕丝
unwind [ʌnˈwaind] *adj.* 松散的,未卷绕的
moth [məθ] *n.* 蛾,蛀虫
cocoon [kəˈkuːn] *n.* 茧,蚕茧
filament [ˈfiləmənt] *n.* 长丝
blend [blend] *v.* 混合
flax [flæks] *n.* 亚麻,麻布
suppress [səˈpres] *v.* 削弱,压制
breed [briːd] *v.* 繁殖,饲养,产生
absorbency [əbˈsɔːbənsi] *n.* 吸收性,吸收率,吸收能力
affinity [əˈfinitiː] *n.* 亲和力
handle [ˈhændl] *n.* 手感
staple [ˈsteipl] *n.* 主要产品,原材料
solely [ˈsəuli] *adv.* 独自地,单独地
mineral [ˈminərəl] *n.* 矿物,无机物
regenerated [riˈdʒenəˌreitid] *adj.* 再生的
synthetic fiber 合成纤维
cellulosic fiber 纤维素纤维
reconstituted [ˌriːˈkənstitjuːtid] *adj.* 再造的,再生的
wood pulp 木纸浆
solvent [ˈsɔlvənt] *n.* 溶剂,溶媒
spinning [ˈspiniŋ] *n.* 纺纱
petroleum [piˈtrəuliəm] *n.* 石油
rayon viscose 粘胶人造丝
artificial silk 人造丝
acetate [ˈæsiˌteit] *n.* 醋酸
lyocell [ˈlīəˌsel] *n.* 天丝,莱塞尔纤维
nylon [ˈnailən] *n.* 尼龙,酰胺纤维;尼龙制品
polyester [ˌpɔliˈestə(r)] *n.* 聚酯
acrylic [əˈkrilik] *adj.* 丙烯酸的
assumption [əˈsʌmpʃən] *n.* 假定;设想
recognition [ˌrekəgˈniʃən] *n.* 认识,识别
whilst [wailst] *conj.* 当……时候

external [eks'tə:nl] *adj*. 外部的，外在的
internal [in'tə:nəl] *adj*. 内部的，内在的
luster ['lʌstə] *n*. 光泽，光彩
prism shape 棱镜形状
cross-section 横截面，剖视图
spinneret ['spinəret] *n*. 吐丝器，喷丝头
bulk [bʌlk] *n*. 膨松度
stiffness ['stifnis] *n*. 硬挺性，硬挺度
resist bending 抗弯曲
resilience [ri'ziljəns] *n*. 回弹，回弹性能
cross-like 十字形的
Y-shaped Y 形的
bean-shaped 豆圆形的

Notes

① Natural fibers come from animal or vegetable sources. All the natural fibers, except cultivated silk, have relatively short fibers which are combed and twisted to form yarn that is strong enough for use in the manufacture of the fabric.
天然纤维来自动植物。除了家蚕丝，所有天然纤维的长度都较短，通过梳织和加捻形成一定强度的纱线，用于织造面料。

② Although frequently blended or woven together, many garments are made entirely from silk, wool, cotton and flax; the other fibers listed are usually mixed with the main fibers to add practical characteristics or aesthetic interest to the fabric.
虽然纱线常被混纺或混织使用，许多服装仍主要由丝、羊毛、棉、麻制成；与这些主要成分混合的其他合成纤维能增强产品的物理性能，并提高面料的美观程度。

③ They are comfortable to wear because of their natural absorbency, and there is great aesthetic appeal in their textures, their dye affinities and their handle.
天然纤维穿着舒适是因为它们与生俱来的吸湿性、美观的织物肌理、良好的染色性和手感。

④ Man-made fibers are produced from chemical solutions that are manufactured into fibers; for example, a chemical liquid can be forced through minute holes and then solidified in air or by chemical processes.
化学方法制造出人造纤维，例如挤压液态化学物质通过小孔，在空气中或通过化学方法使之固化成纤维。

⑤ Originally, man-made fiber lengths were matched to those of existing natural fibers because natural fibers were successful and the new fibers could be processed on existing machinery.
由于天然纤维优良的服用性能，因此，在当时的机械设备允许的情况下，早期的合成纤维都是效仿天然纤维的长度的。

⑥ Until quite recently it was fairly easy to place the fibers of a fabric within a generic group and make certain assumptions about their properties; now, recognition is more difficult.
直到最近，分析面料并对其纤维的性能做出某些推断才变得相对容易一点；现在，纤维的识别更加困难了。

⑦ The shape of the fiber can determine the luster: for example, the filaments of silk are prism shaped and reflect light. The cross-sections of fibers can be changed by varying the holes on the spinneret to match the shape of natural fibers or experiment with new shapes.

纤维的形状决定面料的光泽：例如，蚕丝的长丝结构因为棱形而反光。改变吐丝器的喷孔会导致纤维截面的改变，可以模仿天然纤维的形状，也可以设计出新的形状。

Discussion Questions

1. How the fibers are used?
2. How the fibers are formed into fabrics?

EXTENSIVE READING

NATURAL FIBERS

Cotton

Cotton is a substance that is attached to the seed of a cotton plant. It is grown in 80 countries. Characteristic of cotton include pleasing appearance, high comfort, easy care, durability, and low cost. Prior to 1793, cotton had to be separated from the seeds by hand. In 1793, however, Eli Whitney① invented the saw-tooth cotton gin, which made the processing much faster. Recently, cotton has been mixed with other manufactured textiles, such as spandex, to give the fabric stretch.

图 8 The cotton fiber is bursting out of the pods, cotton is the most widely used natural fiber

Cotton has excellent aesthetic qualities. Its luster is matte, its drape is soft, and its hand is smooth. Cotton is high in comfort in that it has excellent moisture absorbency.

Wool

Wool was one of the first fibers to be woven into cloth. In recent times, wool has been used in place of less expensive substitutions such as acrylic and polyester. There are several types of sheep that produce wool, and there are also several kinds of wool.

图 9 Wool is especially good to wear in odd weather. It keeps body warmth and is very comfortable

Wool has a matte luster. It is high in moisture absorbency and high in thermal retention. It is more difficult to clean than cotton; most wool garments require dry cleaning. Home products made from wool, such as carpet, require steam cleaning.

Silk

Silk is a natural protein fiber. It comes from the cocoon of the silk worm. Two very fine threads come out of an opening in the silkworm's head, and when they are exposed to air, they harden. The four major producers of silk are China, India, Italy, and Japan.

Silk is known for its soft feel, luster, and excellent draping qualities; it is high in comfort because it has high absorbency. Silk wrinkles easily, therefore it has low resiliency.

图 10 The silkworm larvae in their cubicles are at different stages of spinning their cocoons

Linen

图 11 The stems of flax plants furnish long fibers that are made into various weights of linen fabric

Flax is the plant from which linen is made. Flax is twice as strong as cotton. Linen is one of the oldest textiles; in fact, mummies were wrapped in linen.

Aesthetic qualities of linen include a stiff hand. Linen has a low luster and an irregular appearance due to the large variation of thick and thin yarns used to make the fabric. It is high in comfort because of its high moisture absorbency. It has low resiliency, therefore it requires ironing. Generally, linen must be dry-cleaned.

Words and Phrases

easy care 免烫，洗可穿
durability [ˌdjʊrəˈbiliti] *n.* 耐久性，耐用性，坚固
saw-tooth 锯齿形的
cotton gin 轧棉机，轧花机
spandex [ˈspændeks] *n.* 弹力纤维
stretch [stretʃ] *n.* 弹力，弹性，松紧
matte [mæt] *n.* 暗淡色调
moisture absorbency 吸湿性
substitution [ˌsʌbstiˈtjuːʃən] *n.* 代替，替换，代替物
steam cleaning 蒸汽喷洗；蒸汽清洗
thermal retention 保暖性
wrinkle [ˈriŋkl] *n.* 皱褶，折皱
mummy [ˈmʌmiː] *n.* 木乃伊
stiff hand 硬手感
thick and thin 粗细交替的
iron [ˈaiən] *n.* 烙铁，熨斗
linen [ˈlinin] *n.* 亚麻，亚麻制品
dry cleaning 干洗
dry-cleaned 干洗

Notes

① Eli Whitney (1765～1825), American inventor, best known for his invention of the cotton gin.
埃里·惠特尼(1765～1825),美国发明家,因发明轧花机闻名于世。

LESSON 4 FABRICS / 面料

Yarns and fibers are woven, knitted, interlaced or pressed into a fabric form. In most situations it is the fabric which the designer confronts when realizing the design range. Her first aesthetic reactions will be a major part of the criteria that determines the purchase.

Fabric construction can enhance or subdue the characteristics of a yarn. The complex forms that can be produced from the major means of manufacture now offer bewildering choices; textile designers have to balance the visual and textural qualities with its stability and its "fitness for purpose"[①]. It can take a close examination of some fabrics to distinguish which manufacturing process has been used. The bonding of fabrics of different manufacture can confuse simple categorizations.

Fabric Structures

The principal methods of creating fabric are knitting and weaving. The minor methods, interlacing, embroidered and braided are used in many luxury or hand-crafted fabrics. Non-woven fabrics are: felt, many types of interlinings, PVC sheets and some pile fabrics fused onto PVC backing.

Woven Fabric

图 12 Woven Fabric

A fabric is considered to be woven if horizontal threads, the weft, are interlaced with vertical threads, the warp. Garments are usually made up with the wrap threads running down a garment, and the weft threads running across or at an accurate 45 degree angle (which is known as the bias or crossway) to give increased stretch and draping qualities[②].

Pattern pieces are always marked with a grain line to ensure the garment is cut correctly.

The yarns can be interlaced in many different ways to create weaves. Classic weaves become easily recognized: plain weaves give horizontal, vertical and chequered effects; twill weaves give diagonal or herringbone structures; jacquard weaves create complex patterns; satin weaves give smooth surfaces and luster[③]. Different yarns inserted in the warp and weft can give three dimensional rib effects. In pile constructions, yarns in the warp (velvet) or weft (velveteen) give different effects. Many unusual weaves can be created by combining different weaves, or by creating double or double-face fabrics; for example cloque is produced by one set of threads shrinking at a different rate and producing blistering④. Matellasse has extra warp threads inserted which produce a quilted effect. When yarn types and color and print are added to weave constructions, combinations become almost limitless.

Knitted Fabric

- Weft-Knitted Fabric

Weft-knitted fabric is made on machines where the yarn is held by latch needles that move up and down to create rows of interlocking loop stitches across the fabric.

Some machines produce flat fabrics, others circular or tubular fabrics. The structure of the fabric is flexible and varies with the gauge of the machine, the type of yarn, and the tension that the yarn is being held whilst it is knitted. Some machines knit shaped garment parts on fully fashioned knitting machines, but most of the production is produced on flat or circular machines⑤.

图 13 Knitted Fabric

Variations of stitch and patterning: ribbing, inlays, interlock, intarsia and jacquard, create an incredible range of options for the knitwear designer. Weft-knitting can respond to short orders, the machines do not have the complicated setting up of warp threads which is required for woven or warp-knitted fabrics⑥.

- Warp-Knitted Fabric

Warp-knitting machines create vertical interlocking loop chains. Two yarns are often used together to give the fabric stability. There has been a great increase in warp-knitting; the machines are very fast and produce a large amount of fabric from man-made fine filament yarns. The fabrics are particularly suitable for lingerie, openwork and net effects can be produced on the machines.

The names of warp-knit fabrics can be confusing, particularly the pile fabrics; the most common are fleece, terry cloths, velvets, corduroys, loop and pile fabrics⑦. Warp-knit fabrics also provide the backing structure for many laminate and flocked pile fabrics. Depending on its use or its aesthetic effect, the pile side may be used on the face or the back of the fabric.

Words and Phrases

interlace [ˌintəˈleis] *v*. 交织

criteria [kraiˈtiəriə] *n*. 标准

subdue [səb'djuː] *v*. 使屈服，压制，克制；降低

characteristic [ˌkæriktə'ristik] *adj*. 表现特点的，特有的，表示特性的

bewildering [bi'wildəriŋ] *adj*. 令人困惑的，令人不知所措的

stability [stə'biliti] *n*. 稳定性

principal ['prinsəpəl] *adj*. 主要的，最重要的，首要的

bonding ['bɔndiŋ] *n*. 连接，结合，加固，粘合

categorization [ˌkætigəri'zeʃən] *n*. 归类，分类

embroider [em'brɔidə] *v*. 刺绣，在……上绣花

braid [breid] *n*. 辫子，发辫，饰带，编织物，滚带

pile [pail] *n*. 绒毛，绒头，毛茸

pile fabric　绒毛织物，起绒织物，割绒织物

backing ['bækiŋ] *n*. 衬里，底布，背衬

bias ['baiəs] *n*. 斜纹路；斜条；*v*. 斜裁

crossway ['krɔswei] *n*. 斜纹，交叉

pattern ['pætən] *n*. 纸样，裁剪样板；花纹组织；图案；花样，式样

piece [piːs] *n*. 衣片，裁片；部分；匹；件

mark [maːk] *n*. 标记，标志，记号，符号，型号　*v*. 标记，标示

grain line　经向线，经向标志

classic ['klæsik] *n*. 不受时尚影响的，经典的(服装)，传统(服装)

hand-crafted　手工制作的

non-woven　非织物，无纺的

felt [felt] *n*. 毛毯，毡

interlining ['intəlainiŋ] *n*. 衣服衬里，衣服衬里的布料

PVC(polyvinyl chloride)　聚氯乙烯

fuse [fjuːz] *v*. 熔化，使融合，合并

horizontal [ˌhɔri'zɔntəl] *adj*. 地平的，地平线的，水平的

weft [weft] *n*. 纬向，纬纱

warp [wɔːp] *n*. 经向，经纱

vertical ['vəːtikəl] *adj*. 垂直的，竖式的，直立的，纵的

chequered ['tʃekəd] *adj*. 格子花纹的，方格图案的

twill [twil] *n*. 斜纹布，斜纹图案

diagonal [dai'ægənəl] *adj*. 斜的，斜纹的，对角线的

herringbone structure　人字形结构

jacquard [dʒə'kaːd] *n*. 提花机，提花织物

satin ['sætin] *n*. 缎子，绸缎

rib [rib] *n*. 罗纹

velvet ['velvit] *n*. 经绒，丝绒，立绒，天鹅绒

velveteen ['velvi'tiːn] *n*. 纬绒，棉绒，平绒

double-face　双面的

cloque [kləu'kei] *n*. 泡泡纱

shrink [ʃriŋk] *v*. 收缩，缩水

quilted effect　绗缝效果

blister ['blistə] *v*. 起泡

matellasse [mat'ləː sa] *n*. 马特拉塞凸纹布

weft-knitted fabric　纬编针织物

warp-knitted fabric　经编针织物

latch [lætʃ] *v*. 获得，缠住，占有

interlock [ˌintə'lɔk] *v*. 联锁，双罗纹，使连接

loop [luːp] *n*. 线环，布环，毛圈

loop fabric　毛圈织物，毛巾织物

flexible ['fleksəbl] *adj*. 弯曲的，灵活的，弹力的

tension ['tenʃən] *n*. 拉力，张力

shaped [ʃeipt] *adj*. 成型的

fashioned knitting machine　成型针织机

circular ['səːkjulə] *adj*. 圆形的，循环的

tubular ['tuːbjələ] *adj*. 管状的

gauge [geidʒ] *n*. 标准规格，标准尺寸

inlay ['inlei] *v*. 镶嵌，嵌[插]入

intarsia [in'taːsiə] *n*. 嵌花，镶嵌装饰

incredible [in'kredəbl] *adj*. 难以置信的，

不可思议的,惊人的

lingerie [ˌlaːnʒəˈrei] *n*. 女式内衣

openwork [ˈəupənwəːk] *n*. 网状织物,网眼式

net [net] *n*. 网眼织物

confuse [kənˈfjuːz] *adj*. 混乱的,混淆的

fleece [fliːs] *n*. 羊毛,似羊毛物,羊毛标签

terry cloths 毛圈织物

corduroy [ˈkɔːdəˌrɔi] *n*. 灯芯绒

laminates [ˈlæməˌneitz] *n*. 粘合布,多层(粘合)布

flocked fabric 植绒织物

Notes

① Fabric construction can enhance or subdue the characteristics of a yarn. The complex forms that can be produced from the major means of manufacture now offer bewildering choices; textile designers have to balance the visual and textural qualities with its stability and its "fitness for purpose".
织物的组织结构可以加强或削弱纱线的性能。几种主要的织造方法组合起来可以制造出令人眼花缭乱的外观;纺织品设计师常常依据稳定性和实用性对织物的外观和性能进行取舍。

② A fabric is considered to be woven if horizontal threads, the weft, are interlaced with vertical threads, the warp. Garments are usually made up with the wrap threads running down a garment, and the weft threads running across or at an accurate 45 degree angle (which is known as the bias or crossway) to give increased stretch and draping qualities.
水平方向的纬纱和垂直方向的经纱交织而成的面料称为梭织面料。服装通常沿经向裁剪并制作完成。水平纬向或倾斜 45°方向(称为斜丝缕或斜向)的织物具有较高的伸缩性和悬垂性。

③ Classic weaves become easily recognized: plain weaves give horizontal, vertical and chequered effects; twill weaves give diagonal or herringbone structures; jacquard weaves create complex patterns; satin weaves give smooth surfaces and luster.
传统的织造法容易理解:平纹组织产生水平、竖条和格子的效果;斜纹组织产生斜向或人字形组织结构;提花组织可织造出复杂的花样;缎纹组织表面滑爽并富有光泽。

④ In pile constructions, yarns in the warp (velvet) or weft (velveteen) give different effects. Many unusual weaves can be created by combining different weaves, or by creating double or double-face fabrics; for example cloque is produced by one set of threads shrinking at a different rate and producing blistering.
起绒面料的组织结构中,处理经向(经绒)和纬向(纬绒)的纱线会产生不同的效果。组合不同的织造法,或者织造双层或双面面料都会产生许多新奇的组织结构。例如泡泡纱,就是对组织中的一组纱线不同程度皱缩处理而产生的起泡效果。

⑤ Some machines produce flat fabrics, others circular or tubular fabrics. The structure of the fabric is flexible and varies with the gauge of the machine, the type of yarn, and the tension that the yarn is being held whilst it is knitted. Some machines knit shaped garment parts on fully fashioned knitting machines, but most of the production is produced on flat or circular machines.

有些纬编机生产平面面料，有些生产圆形或筒形面料。纬编面料具有弹性，并随着纬编机器型号、纱线种类和纺纱张力的变化而变化。成型的针织机直接织造出成型的衣片，但大多数针织产品是在平面或圆形纬编机上制造的。

⑥ Weft-knitting can respond to short orders, the machines do not have the complicated setting up of warp threads which is required for woven or warp-knitted fabrics.

纬编织法可以用于短期订单，因为没有梭织机或经编机那些复杂的经线控制设置。

⑦ The names of warp-knit fabrics can be confusing, particularly the pile fabrics; the most common are fleece, terry cloths, velvets, corduroys, loop and pile fabrics. Warp-knit fabrics also provide the backing structure for many laminate and flocked pile fabrics.

经编针织物的称谓容易混淆，特别是起绒织物，常见的有：羊绒、毛圈绒、经绒、灯芯绒、圈圈绒等。经编面料还是很多粘合布和植绒面料的底布。

Discussion Questions

1. Explain how fabric characteristics affect the fabric performance.
2. Evaluate fabrics for use in clothing, home furnishings, and recreational items.

EXTENSIVE READING

FABRIC FINISHES

All fabric are finished , the simplest form is simply washing, shrinking and pressing, but most fabrics have some form of extra finish, many of which are complex and may be a completion of an earlier process of manufacture. The finish may be added to improve the aesthetic and tactile quality of a fabric, to enhance or suppress its natural properties, or to add some specific or novel quality. The finish can be permanent or temporary.

White dyed or printed fabrics produced from cellulose fibers have to be bleached. Fabrics made from fibers that have a rough texture can be smoothed by cropping and singeing, be chemical finishes, or calendaring, glazing and engraving. They can have their rough appearance enhanced; the surfaced is raised by brushing or plucking. Many of these fabrics then have a pile or "nap" that is usually cut one way with the fibers lying towards the bottom of the garment; However, interesting effects can be created by cutting up and down a napped fabric. Thicker and softer yarns woven on the back of a fabric can be brushed to give an outer flat appearance and a warm fleecy back. Stripe effects can be made by pile finishes on groups of warp yarns.

Fabrics made from filament yarns are usually smooth and lustrous, and many of those made from man-made fibers imitated silk. However, experiments with combinations of fiber and yarn structures and finishes have created a large explosion of new fabrics which respond to unusual finishes and produce unique combinations of qualities.

Combinations of many of the above techniques, in particular areas of finishes targeted at particular yarns, can give uneven and sculptural effects to fabrics. Fabrics made from thermoplastic synthetic yarns can achieve similar effects by heat-setting. Coating of bonding fabrics usually produces dramatic change to any fabric; many of the coatings on bonding are thermoplastic and are heat-set.

Most of the processes discussed are those that alter the characteristics that are of interest to the process of pattern cutting. However, many finishes are developed for garments that are used for particular purposes, for example athletic wear requires high absorbency, weather-wear requires waterproofing, and some industrial wear requires chemical and flame proofing. The designer has to consider this kind of parameter when developing a range in a particular product field.

图 14 Fireman's overall with fireproof performance

Some finishes are completed after the garment is made up. This means that quite complex shrink allowances are required during the development of the pattern, and tight controls are required on the finishing processes.

Words and Phrases

finish [ˈfiniʃ] *n*. 织物整理,后整理
washing [ˈwɔʃiŋ] *n*. 洗涤
pressing [ˈpresiŋ] *n*. 熨烫,压呢,烫衣
tactile [ˈtæktəl] *n*. 手感,触感
novel [ˈnɔvəl] *adj*. 新奇的,新颖的
permanent [ˈpəːmənənt] *adj*. 永久的
temporary [ˈtempərəri] *adj*. 暂时的
white dye 增白,加白
cellulose [ˈseljuləus] *n*. 纤维素
bleach [bliːtʃ] *v*. 漂白,褪色
rough [rʌf] *adj*. 布面毛糙的
smooth [smuːð] *adj*. 光滑的,滑爽的
cropping [ˈkrɔpiŋ] *v*. 剪除,剪去
singe [sindʒ] *v*. 烧去(布匹的)茸毛
calender [ˈkælində] *n*. 轧光整理
glaze [gleiz] *v*. 上光,极光;光泽,色泽
engrave [inˈgreiv] *v*. 雕刻,刻印
brush [brʌʃ] *v*. 擦,刷亮
pluck [ˈplʌk] *v*. 拔毛,拉毛
nap [næp] *n*. 衣料起毛 *v*. 使衣料起毛
lay [lei] *v*. 铺放,铺料
bottom [ˈbɔtəm] *n*. (织物)底色,下方;(衣服)下摆,下装
cutting [ˈkʌtiŋ] *n*. 裁剪
up and down 上上下下的
napped fabric 起绒面料
outer [ˈautə] *adj*. 外在的,外部的
appearance [əˈpiərəns] *n*. 外观,外貌
back [bæk] *n*. (织物)背面
fleecy [ˈfliːsi] *adj*. 蓬松的,羊毛似的
stripe [straip] *n*. 条子,条纹
lustrous [ˈlʌstrəs] *adj*. 光亮的,鲜艳的
uneven [ʌnˈiːvən] *adj*. 不均匀的
heat-setting 热定型
coating [ˈkəʊtiŋ] *n*. 上胶,涂层,涂料
thermoplastic [ˌθəːməˈplæstik] *adj*. 热塑性的 *n*. 热塑性塑料
weather-wear 风雨衣

waterproofing [ˈwɔːtəpruːfiŋ] *n.* 防水，绝湿

flame proofing 抗燃，阻燃，防火

parameter [pəˈræmitə] *n.* 参数，参量

LESSON 5 FABRIC PERFORMANCE / 面料性能

Fabric performance can be divided into three areas: durability, comfort, and ease of care.

Durability

Durability refers to all those characteristics that affect how long you will be able to wear or to use a particular garment or item. These include strength, shape retention, resiliency, abrasion resistance, and colorfastness①.

● Strength

Is the fabric going to be strong enough for the way that you plan to use it? Different fibers have different tensile strengths, or ability to withstand tension or pulling. Strength is also related to the fabric construction. Tightly woven or knitted fabrics are usually stronger than loosely woven or knitted fabrics. For example, canvas is used as chair seats, while a sheer open-weave fabric is used for curtains②.

● Shape Retention

Will the fabric retain its shape after wearing or cleaning, or will it stretch so that you end up with baggy knees and elbows? Shape can also be lost when the fabric is washed or machine-dried. Some fibers shrink when exposed to water or heat.

● Resiliency

Is the fabric resilient, or able to spring or bounce back into shape after crushing or wrinkling? Will the wrinkles hang out of the garment, or must the fabric be pressed? The fibers in a wool carpet may flatten underneath a piece of heavy furniture. However, they will spring back into shape when the carpet is steamed.

● Abrasion Resistance

Will the fabric resist abrasion? Abrasion is a worn spot that can develop when the fabric rubs against something. This can occur on the inside of a collar where it rubs the back of your neck or at your side where you carry your books. Some fabrics can pill, or form tiny balls of fiber on the fabric.

● Color-Fastness

Color-fastness means that the color in the fabric will not change. It will not fade from washing, from chlorine in a pool, or from exposure to sunlight③. However, some denim blue jeans are meant to lighten when washed. Madras, a woven plaid fabric, is meant to bleed so that the plaid becomes softer and less distinct.

Comfort

Comfort is another factor to consider when selecting fabric. A fabric can be the right weight and texture, durable and easy to care for, but uncomfortable to wear. It may be too hot, too cold, or too clammy. A fabric's absorbency, wicking ability, breathability, and stretchability all affect how comfortable the fabric is on your body.

● Absorbency

This term refers to how well the fabric takes in moisture. Some fibers, such as cotton and wool, are very absorbent. Other fiber, such as polyester and nylon, are not. That is why you may feel very clammy when you wear 100% polyester fabric on a hot summer day④. Your perspiration stays on the surface of your skin and is not absorbed by the fabric. That is also why you can dry yourself faster with a terry cloth towel made from 100% cotton than with one made from a cotton and polyester blend. However, special finishes can be applied to improve the absorbency of fabrics.

图 15 Sportwear should be comfort and good absorbency

● Wicking

This term refers to a fabric's ability to draw moisture away from your body so that the moisture can evaporate. The wicking ability of some fibers makes up for the fact that they are not very absorbent. Olefin, a fabric that you will be learning more about, has wicking properties.

● Breathability

This characteristic is another important factor to consider when choosing comfortable fabrics. It refers to the ability for air or moisture to pass through fabric. Some fabrics have special finished to prevent rain and moisture from penetrating the fabric. These finishes also prevent body moisture from evaporating through the fabric. That is why your feet and your body often sweat when you wear rubber boots and a rubber raincoat⑤. Manufacturers sometimes compensate for this in waterproof clothes by adding small grommet holes under the arms to act as air vents.

● Stretchability

This term describes the fabric's ability to "give" and stretch with the body. How much stretchability you will need in your clothes depends on your activities. You might want extra stretchability in your swimsuit, skipants, and exercise wear.

Ease of Care

The type of care that a fabric requires determines how easy it is to care for a garment or other item. Washing, dry cleaning, ironing, brushing, and folding are all methods of fabric care⑥.

Some fabrics require more routine care than others. When selecting fabrics, you should choose those that match your lifestyle. Washability, soil and stain resistance, and wrinkle resistance are some factors that influence fabric care.

- Washing Ability

Can the fabric be washed or must it be dry cleaned? Over a long period of time, your dry cleaning bills may add up to more than the cost of the garment. Do you have the time and space to hand wash and dry different garments? Although it is easy to hand wash a sweater, will you set aside the time to do it? Will the fabrics shrink more than one or two percent? If so, it might affect the fit of the garment.

- Stain and Spot Resistance

Is the fabric resistant to stains and spots? Some fibers absorb stains, but special finishes can help the fabric to repel the stain or to release it during cleaning. Carpets, upholstery, coats and jackets, and children's clothes often have special finishes⑦.

- Wrinkle Resistance

Do you have to iron the fabric every time that it is washed? Do you have to press a garment before every wearing, or will the fabric wrinkles hang out after a short time? Fibers have different characteristics that affect wrinkling. For example, polyester is very wrinkle-resistant, but cotton and rayon wrinkle easily. Special finishes, such as durable press, can be applied to fabrics to improve their wrinkle resistance⑧.

Words and Phrases

comfort [ˈkʌmfət] *n*. 舒适,舒适性
ease of care 易护理性
strength [streŋθ] *n*. 强度
shape retention 保形性
abrasion resistance 抗磨损性
colorfastness [ˈkʌləfaːstnis] *n*. 染色坚牢度
tensile strength 抗张强度
canvas [ˈkænvəs] *n*. 帆布
sheer [ʃiə] *n*. 透明薄织物;透明薄纱
open-weave 稀薄组织
elbow [ˈelbəu] *n*. 肘部
resiliency [riˈziliənsi] *n*. 跳回,弹性
spring [spriŋ] *v*. 弹回,反弹
bounce back 迅速恢复活力
crush [krʌʃ] *v*. 压皱,揉皱
flatten [ˈflætn] *v*. 拉平
underneath [ˌʌndəˈniːθ] *v*. 在下面,在……的下面
rub [rʌb] *v*. (织物的)擦伤痕
pill [pil] *v*. 起球
fade [feid] *v*. 褪色,枯萎;凋谢
chlorine [ˈklɔːriːn] *n*. 氯
exposure [iksˈpəuʒə] *v*. 暴露,揭发,揭露
denim [ˈdenim] *n*. 斜纹粗棉布,牛仔布,劳动布
madras [məˈdrɑːs] *n*. 马德拉斯狭条衬衫布
plaid [plæd] *n*. 方格布,方格呢;格子花纹
bleed [bliːd] *v*. 渗出;(印染等)渗色,渗开
distinct [disˈtiŋkt] *adj*. 独特的;明显的;不寻常的
clammy [ˈklæmiː] *adj*. 滑腻的,粘糊糊的,冷淡的
wicking ability 吸附能力
stretchability [ˌstretʃəˈbiliti] *n*. 拉伸性;延伸性;拉伸性
moisture [ˈmɔistʃə] *n*. 湿气,水分,潮湿;水蒸气
perspiration [ˌpəːspəˈreiʃən] *n*. 汗(水);出汗

terry cloth （毛巾、围巾等）两端留有绒穗的物品
evaporate [iˈvæpəreit] v. 使蒸发；使脱水
olefin [ˈəuləfin] n. 烯烃
breathability [breθəˈbiliti] n. 透气性
penetrate [ˈpenitreit] v. 穿透，穿过，透过
rubber boots 橡胶靴，橡胶长统靴
raincoat [ˈreinkəut] n. 雨衣，风雨衣
grommet holes 金属孔眼，索环
arm [aːm] n. 袖子；手臂，上肢
air vents 通风口，透气孔
swimsuit [ˈswimˌsuːt] n. 女游泳衣
skipants [skiːpænts] n. 滑雪裤
ironing [ˈaiəniŋ] n. 熨烫
folding [ˈfəuldiŋ] n. 折叠
hand wash 手洗
washability [wɔʃəˈbiliti] n. 可洗性
stain and spot resistance 防沾污性
upholstery [ʌpˈhəulstəriː] n. 室[车]内装饰（品）
wrinkle-resistant 防皱处理，耐褶皱性
wrinkle resistancy 抗皱性，抗皱处理
rayon [ˈreiən] n. 人造丝织物，人造丝
durable press 耐久压烫

Notes

① Durability refers to all those characteristics that affect how long you will be able to wear or to use a particular garment or item. These include strength, shape retention, resiliency, abrasion resistance, and colorfastness.
耐久性是指所有影响一件服装使用寿命或某个功能的性能。包括面料的强度、形状保持性、回复性、抗磨损性和色牢度。

② Is the fabric going to be strong enough for the way that you plan to use it? Different fibers have different tensile strengths, or ability to withstand tension or pulling. Strength is also related to the fabric construction. Tightly woven or knitted fabrics are usually stronger than loosely woven or knitted fabrics. For example, canvas is used as chair seats, while a sheer open-weave fabric is used for curtains.
面料的强度能满足使用中的需要么？不同的纤维对外界的拉扯具有不同的抗拉强度，面料的强度也与面料的组织结构有关。质地紧密的梭织、针织物通常较质地稀松的织物强度大。例如，帆布可用于缝制座椅，而薄型网眼布只能用于缝制窗帘。

③ Color-fastness means that the color in the fabric will not change. It will not fade from washing, from chlorine in a pool, or from exposure to sunlight.
染色牢度是指面料上颜色的固色程度，不会由于水洗、氯的漂白或日晒而褪色。

④ This term refers to how well the fabric takes in moisture. Some fibers, such as cotton and wool, are very absorbent. Other fiber, such as polyester and nylon, are not. That is why you may feel very clammy when you wear 100% polyester fabric on a hot summer day.
这个性能是指面料对水分的吸收程度。有些纤维，例如棉和羊毛，吸湿性良好，其他的纤维，例如涤纶和尼龙，吸湿性较差。这就是为何大热天穿一件100%涤纶的衣服，你会感觉非常湿热的原因。

⑤ This characteristic is another important factor to consider when choosing comfortable fabrics. It refers to the ability for air or moisture to pass through fabric. Some fabrics have special finished to prevent rain and moisture from penetrating the fabric. These finishes also prevent body moisture from evaporating through the fabric. That is why your feet and your body often sweat when you wear rubber boots and a rubber raincoat.

这是选择舒适面料时需要考虑的另外一个重要因素。它是指面料对水、气的通透性。有些面料经过专门的后整理具备防雨、防潮的功能，这些后整理也阻碍了人体表面水分的挥发，这就是当你穿橡胶鞋子和雨衣时常会感觉出汗的原因。

⑥ The type of care that a fabric requires determines how easy it is to care for a garment or other item. Washing, dry cleaning, ironing, brushing, and folding are all methods of fabric care.

面料具有这种性能意味着服装或服饰易于护理：例如方便洗涤、干洗、熨烫、刷毛和折叠等所有面料的相关护理。

⑦ Is the fabric resistant to stains and spots? Some fibers absorb stains, but special finishes can help the fabric to repel the stain or to release it during cleaning. Carpets, upholstery, coats and jackets, and children's clothes often have special finishes.

面料能防污、耐污么？有些纤维会吸收污迹，但是专门的后整理可以帮助面料耐污和去污。地毯、室内软装饰、大衣和茄克，还有儿童服装常常需要这样专门的后整理。

⑧ Fibers have different characteristics that affect wrinkling. For example, polyester is very wrinkle-resistant, but cotton and rayon wrinkle easily. Special finishes, such as durable press, can be applied to fabrics to improve their wrinkle resistance.

纤维不同的性能影响面料的皱缩性。例如，涤纶抗皱性良好，但棉和人造丝则易折皱。特殊的后整理，例如耐久压烫，可以用来改善面料的抗皱性。

Discussion Questions

1. Describe the different characteristics of fabric.
2. Explain why the knowledge of fabric can help you to make wise consumer decisions.

EXTENSIVE READING

FABRIC PRODUCTION

Fabric Widths

Fabrics are produced in piece lengths which vary in length and width. The piece length is decided by the weight and bulkiness of the fabric. The width will vary from 72cm to circular jersey fabric which can be as wide as 180cm. Light-weight fabrics have mainly been woven 90－114cm in width; but companies, who produce large quantities, are demanding wider fabrics to gain greater efficiency in their garment lays. Woolen fabrics and tweeds are generally woven at 150cm width. The width of the fabric is crucial to the garment designer. Costing negotiations frequently require modifications to a design, the final cut of the garment may be determined by the width of the fabric.

Fabric Weights

Fabric weights are given in two ways: weight per running meter or weight per square meter. The latter is the most useful when comparing different qualities. Fabric swatches do not always state the type of weight; therefore the designer or technologist may have to re-weigh a sample piece. Very light-weight fabrics have to be made from strong fibers or specially processed fibers and tend to be more expensive than medium-weight fabrics. Heavy-weight fabrics are usually expensive because of the quantity of yarn used. Exceptions to these generalizations can be found and hard wear or strength may not be characteristics of principal concern for the design.

Fabric Thickness

The thickness of a fabric is dependent on a large number of variables; the fiber structure, the yarn structure and finish, the fabric structure and finish, surface decoration, fabric bonding or lamination. Double-faced fabrics can be made by interweaving two layers of woven cloth, or in knitting, using the front and back needles. A great improvement in bonding techniques has led to many combinations of fabrics being bonded: to give strength to a weak of flexible structure; to "sandwich" insulating fabric to create reversible cloths; to bond weatherproof membranes; to create a particular handle or three dimensional appearance.

图 16 Each type of fabric has its width, weight and thickness

Words and Phrases

width [widθ] *n.* 幅宽
bulkiness [bʌlkinis] *n.* 膨松性，膨松度
light-weight 轻量
medium-weight 中厚
heavy-weight 厚重
tweed [twiːd] *n.* 粗花呢
modification [ˌmɔdəfiˈkeiʃən] *n.* 修改，修正
technologist [tekˈnɔlədʒist] *n.* 工艺师
hard wear 经穿，耐磨
variable [ˈvɛəriəbl] *n.* 变数，变化因素
crucial [ˈkruːʃəl] *adj.* 决定性的；关键的，重要的
negotiation [niˌgəuʃiːˈeiʃən] *n.* 协定，交易，协商，谈判
latter [ˈlætə] *adj.* 后者的
swatch [swɔtʃ] *n.* 小块样布；样品，样本
lamination [ˌlæmiˈneiʃən] *n.* 粘合衬
interweave [ˌintəˈwiːv] *v.* 交织，混合
insulate [ˈinsjuleit] *v.* 隔离，使孤立
reversible [riˈvəːsəbl] *n.* 双面织物
weatherproof [ˈweðəpruːf] *adj.* 防风雨的
membrane [ˈmemˌbrein] *n.* 膜，薄膜

LESSON 6 FINDINGS AND TRIMS / 服装附件

图 17 Findings and trims

Findings and trims are called upon to perform either functional or decorative roles, but some may do both. Functional and trims are an integral part of the garment structure.

Findings and trims must be compatible with the outer shell fabrics during wear and maintenance. The performance standards for them should be based on the use, care, and construction of outer shell fabrics for the style①.

Findings

- Interlining. An extra lining inserted between the fabric and lining of a curtain or piece of clothing to make it thicker or warmer, or the fabric used for this②.
- Lining. Either real silk, artificial silk or cotton. Linings for ladies' garments are usually 90—100cm wide. Gentlemen's linings are always 140—150cm wide.
- Interfacing. A kind of stiffening fabric used to stiffen or support collars, cuffs, or other parts of a garment.
- Canvas. Canvas for interlinings. The best type is made of fine linen, and should be a suitable weight to the cloth. Cotton canvas is used for unimportant interlining such as cuffs, etc. Wool and hair canvas is very suitable for heavy garments.
- Stay Linen. Stay linen is a particularly strong, soft, thin linen used for strengthening pockets, buttonholes, coat edges, etc. it can be obtained in black, browns and grays③.
- Silesia. A heavily "milled" cotton material used for pocket bags in coats. Most colors are available but it is mostly used in black, browns, grays and white④.
- Stay Tape. Stay tape made of linen and used for strengthening the edge of coats, pocket mouths, etc. in 1cm width.
- Wadding. Used for padding shoulders and sleeve head seams, in black and white.
- Shoulder Pads. Shoulder pads are used in some garments, and made in various sizes and thicknesses and of different types of materials such as wool, felt, sponge rubber, etc.
- Glissade. Glissade is a special sleeve lining for heavy over-garments. It is slippery in two directions, which enables the coat to slip on and off very easily⑤.
- Machine Cotton and Silk. A smooth, fine cotton or silk with a left-hand twist may be obtained in various thicknesses in reels and cops.
- Basting Cotton. A rough heavy white cotton, easily broken, in reels or cops⑥.
- Hank Silk. A fine, strong silk for hand sewing available in three main colors, blue-black, browns and grays.

- Zippers. A fastener for clothes, bags, or garments consisting of two rows of interlocking metal or plastic teeth with an attached sliding tab pulled to open or close the fastener.
- Button. A flat and usually round piece of plastic or other material on a piece of clothing that fits into a slit or loops on another part and holds the two parts together.
- Hooks and Eyes. A fastening for clothes consisting of a small hook inserted into a metal or thread loop.
- Snap. A circular fastener consisting of two halves that close when pressed together and open when pulled apart.
- Thread. Fine cord made of two or more twisted fibers. Use: sewing, weaving.
- Linen Thread. In hanks and reels. Hank thread is the stronger, and is used for buttons and other strong sewing jobs, also for strengthening buttonholes.
- Elastics. Elastic is a kind of strips or threads of rubber or similar stretchable material, or fabric or tape with a stretchy material woven into it so that it can fit tightly around something.
- Label. Label is a piece of paper, fabric, or plastic attached to garment to give instructions about it or identify it.

Trims

- Linear Trims. Linear trims from lines on the surface or edges of the garment. Decorative edgings and seams are forms of linear trims.
- Narrow Fabric Trims.
- Ribbon. Ribbon is a narrow woven fabric with finished edges. It is available in width of 1/8 inch to several inches. Similar to other fabrics, vary significantly in quality.
- Passementerie. Passementerie is an umbrella term that includes a broad range of braids and cords in straight, curved, fringe, or tassel forms[7].
- Lace. Lace is an openwork trim or fabric that is made into intricate designs by the intertwining of many threads. Lace can be light and airy or heavily textured.

Words and Phrases

findings [ˈfaidiŋz] *n*. 服装附件
integral [ˈintigrəl] *adj*. 组成的,必备的
compatible with 协调的,相容的
shell fabric 面料
maintenance [ˈmeintinəns] *v*. 维持,保持
inter lining [ˈintəˈlainiŋ] *n*. 衣服衬里;衣服衬里的布料
interfacing [ˈintəːfeisiŋ] *n*. 粘合衬
stiffen [ˈstifn] *vt*. 使硬,使粘稠
pocket [ˈpɔkit] *n*. 口袋
buttonhole [ˈbʌtnhəul] *n*. 扣眼
Silesia [saiˈliːzjə] *n*. 西里西亚里子布
mill [mil] *v*. 缩绒,缩呢
stay tape 定位带,胸衬条,牵条,过桥布
pocket mouth 袋口
wadding [ˈwɔdiŋ] *n*. 填絮,软填料,衬垫
sleeve head 袖山头
shoulder pads 垫肩,肩衬,护肩
thickness [ˈθiknis] *n*. 厚度,密度
sponge rubber 橡胶海绵,多孔橡胶,泡沫

橡胶

Glissade [gli'sa:d] *n.* 格利萨特里子布

over-garment 大衣,罩袍

slippery ['slipəri] *adj.* 易脱落的,手感滑的

slip [slip] *n.* 活络里子,套裙,女式长衬裙,儿童围兜

machine cotton and silk 缝纫机用棉线,丝线

reel [ri:l] *n.* 线轴

cop [kɔp] *n.* 纺锤状线团

basting cotton 粗缝棉线,绗缝棉线

button-hole twist 锁钮孔线

hank silk 丝线团

zipper ['zipə] *n.* 拉链

fastener ['fa:sənə] *n.* 紧固件

slide tab 拉链头子,拉链滑块

button ['bʌtən] *n.* 扣子,钮扣,按钮

slit [slit] *n.* 缝,裂口,槽; *v.* 开衩

hooks and eyes 钩眼扣子;风钩

thread loop 线环

snap [snæp] *n.* 按钮,钩扣

thread [θred] *n.* 线

cord [kɔ:d] *n.* 线绳,滚条

linen thread 麻线

elastic [i'læstik] *n.* 弹松带,松紧带,橡皮带;弹性织物

stretchable [stretʃəbl] *adj.* 有弹性的

tape [teip] *n.* 狭幅织物,狭带,卷尺 *v.* 贴边,镶边,牵条

label ['leibəl] *n.* 标号,标签,标记,商标;服装商店(或时装设计师)的标记

linear trim 线形饰边

ribbon ['ribən] *n.* 带,丝带,缎带,饰带

passementerie [pas'mətri:] *n.* 边饰,金银线镶边,珠饰

straight [streit] *adj.* 直线的;直筒的

curved [kə:vd] *adj.* 曲线的

fringe [frindʒ] *n.* 毛边,蓬边;流苏

tassel ['tæsəl] *n.* 穗,流苏

lace [leis] *n.* 网眼织物,网眼花边,滚带,饰带

intricate ['intrikit] *adj.* 复杂的,缠结的

airy ['eəri:] *adj.* 透气的

Notes

① Findings and trims must be compatible with the outer shell fabrics during wear and maintenance. The performance standards for them should be based on the use, care, and construction of outer shell fabrics for the style.
配饰件必须与面料在服用性能和维护上保持一致。如何选择配饰件可根据面料的种类、日常护理以及款式结构来定。

② Interlining. An extra lining inserted between the fabric and lining of a curtain or piece of clothing to make it thicker or warmer, or the fabric used for this.
衬布:是针对服装某个部位加固的衬料,用于面料和里料之间,使得这个部位更加牢固或保暖。

③ Stay Linen. Stay linen is a particularly strong, soft, thin linen used for strengthening pockets, buttonholes, coat edges, etc. it can be obtained in black, browns and grays.
加固麻布条:特别结实的、柔软的薄型麻布。用于加固口袋、钮扣和大衣下摆等位置。可用的有黑色、棕色和灰色。

④ Silesia. A heavily "milled" cotton material used for pocket bags in coats. Most colors are available but it is mostly used in black, browns, grays and white.

西里西亚里子布：由缩绒的棉材料制成，用于大衣口袋的缝制。可选颜色很多，常用色有黑色、棕色、灰色和白色。

⑤ Glissade. Glissade is a special sleeve lining for heavy over-garments. It is slippery in two directions, which enables the coat to slip on and off very easily.
格利萨特里子布：大衣袖子的里布，表面光滑，易穿脱。

⑥ Basting Cotton. A rough heavy white cotton, easily broken, in reels or cops.
粗棉缝线：一种粗糙的白色棉线，易拉断，有线轴线团，也有纺锤线团。

⑦ Passementerie. Passementerie is an umbrella term that includes a broad range of braids and cords in straight, curved, fringe, or tassel forms.
边饰是指一系列直条的、弧形的、流苏的或穗状的镶边和滚条的统称。

Discussion Questions

1. Which findings and trims would be required to complete the entire line of garments you designed?
2. What criteria are used to select interfacings and trims?

EXTENSIVE READING

BUTTON

The buying public can often underestimate the importance of trimmings and accessories to the fashion world. The button industry, for instance, is large and sophisticated and involves research, design, highly skilled manufacture and distribution. Its markets are wholesale, retail and high fashion. The process starts from the raw materials; buttons can be made from both natural and man-made materials and a mixture of both. Bone, horn, tortoiseshell, mother of pearl, glass and metal were the traditional materials, but they have been superseded by nylon, plastic and polyurethane. Until quite recently the bulk of good quality buttons were made from casein, a hardened by-product of milk, with the qualities of bone which could be carved and molded in a similar way.

Button manufacturers may specialize in a particular field: metal buttons for military uniforms, fashion buttons, practical heavy-duty buttons for work-wear and other specialist tasks. The fashion button is, however, the most common, and companies producing this type of button will offer hundreds of styles with many new designs created for each season. Button manufacturers show their styles at fabric, fashion and product fairs so that the designers and clothing manufacturers can choose button styles for the coming season.

Button designs are shown on button cards and each card will represent a particular look-classics like tortoiseshell and pearl, fancy shapes, together with buckles, clips and toggles. Most cards will have between 10 and 20 styles and companies will normally show 30 to 40 cards. Buttons have a method of sizing all the measurements in lignes (14-100);

图 18 Chinese frog

a one inch diameter button is, for example, 40 lignes. Designers such as Chanel and Saint Laurent① have always had exclusive buttons, which have almost become part of their trademark.

Good designers will have prepared the way before starting to design a collection. Not only will they have attended fabric fairs, ordered first cloth samples, become aware of any color predictions and forecasts but they will also have made sure that they have seen all the available button cards showing next season's button and buckle samples, belt designers' samples, embroiderers' watches and new trimming notions.

Words and Phrases

underestimate [ˌʌndər'estimeit] *v*. 低估，估计不足
mixture ['mikstʃə] *v*. 混合，混合状态；*n*. 混合物
tortoiseshell ['tɔːtəʃel] *n*. 龟甲
supersede [ˌsjuːpə'siːd] *v*. 代替
nylon ['nailən] *n*.
polyurethane [ˌpɔli'juəriθein] *n*. 聚氨基甲酸脂；聚氨酯(类)
casein ['keisiːin] *n*. 酪朊，酪蛋白；酪素
hardened ['haːdənd] *adj*. 坚固的，坚硬的
by-product 副产品
carve [kaːv] *v*. 切割，雕刻
heavy-duty 耐用的，重型的
buckle ['bʌkl] *n*. 扣子，带扣
clip [klip] *n*. 别针，首饰别针
toggle ['tɔgl] *n*. 套环，套索扣
ligne 莱尼(钮扣规格：1 莱尼＝0.633 毫米)
diameter [dai'æmitə] *n*. 直径
trademark [treidmaːk] *n*. 商标，牌号
embroiderer [em'brɔidərə] *n*. 绣花机；绣花工
Chinese frog 中国盘扣

Notes

① Yves Saint Laurent (1936～2008), French fashion designer, whose work had an exceptional influence on fashion in the second half of the 20th century.
伊夫·圣·洛朗(1936～2008)，20 世纪下半叶极具影响力的法国著名时装设计师。

Chapter 3 FASHION DESIGN / 服饰设计

LESSON 7 TEXTURE, FABRIC PATTERNS AND COLOR / 肌理、图案和色彩

Texture

Texture, a visual characteristic that affects the hand of the fabric, is a result of the way the textile is constructed, the fiber and yarn used, and the way the fabric is finished.

Draping fabric on the dress form and sewing fabric to create garments teach a person how textiles react in various silhouettes. Converting textiles, flat two-dimensional material, into apparel provides essential training for the commercial designer exactly as working with clay and molding it into plates, bowls, and other items teaches the potter his craft①.

Shiny fabric reflects light and visually enlarges the area it covers, especially when a garment fits snugly. Fabric with a low luster reflects the light to a lesser degree than a shiny fabric and subtly enhances the wearer's face②. Silk crepe is a typical low-luster fabric. Matte fabric has no luster, does not highlight the area it covers, and is a camouflage especially when dyed a neutral or dark color.

Crisp, absorbent fabrics, like cotton and linen, are appropriate for summer clothing because they absorb perspiration and allow the body to breath. Spongy fabrics with a soft and lofty hand feel warm because the thickness of the fabric retains the body's heat③. Thick woolens and quilted fabrics are appropriate for cold weather outerwear because they insulate by keeping the body's heat from escaping.

Observe and experiment with fabric to learn how the intrinsic characteristics of textiles translate into garments which are appropriate for various practical end uses and styles.

Fabric Patterns

Pattern designs on fabric are created by color, lines, shapes, and spaces. They come in an endless variety-stripe, plaids, geometrics, floras, scenic, borders, and many others. The designs can be large or small, even or uneven, light or dark, spaced or clustered, muted or bold. All will affect how a fabric will look on you④.

Fabric patterns, just like fabric texture, can create illusions in design. Small prints in subdued colors usually decrease apparent size. Large, overall designs increase size. Widely spaced motifs will also make you seem larger. Prints with large curves give a feeling of added roundness and size.

When selecting striped or plaid fabrics, you should follow the basic theory of line and illusion. Identify the dominant stripe of the fabric. If you cannot identify it easily, try the squint test. As you squint at the fabric, note which stripe you see first. This will be the dominant one⑤. Placement of this stripe is important. How will it divide up the space within your garment? Is it vertical, horizontal, or diagonal?

Where on the body do the stripes fall? For example, a bold horizontal stripe across the hipline or waistline will make either look larger. What happens when the stripes meet at the seams? Do they match horizontally or vertically? Do they chevron, or form angles, at the seams?

Select prints, stripes, and plaids that are in scale, or in proportional size, with your own body size. Small designs look best on the small to average person, but they look out of place and lost on a large figure. On the other hand, large designs are best worn by the average to tall person as these designs can overwhelm a small figure⑥.

Color

It has been found that the first element to grab the customer's eye is the color of a garment. Color is the most fundamental fashion element; it is usually the first decision a designer makes in each season.

- Hue

The difference between one color (red, blue) and another (yellow).

- Chroma

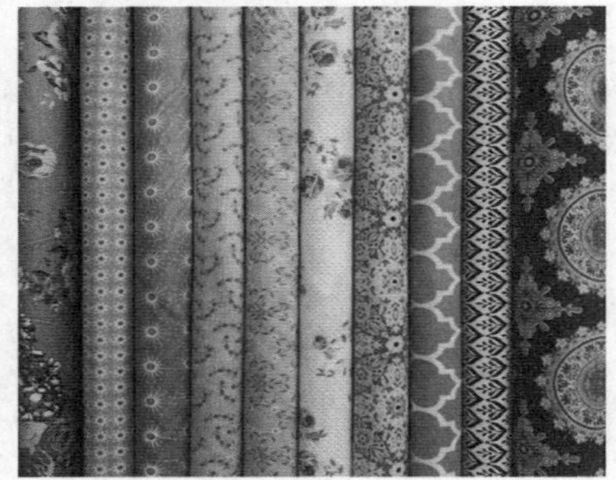

图 19 Fabric with patterns

Saturation or intensity of a particular color-the difference between brightness and dullness. Stated another way, it is the amount of gray in the color. For example, a dusty blue and a very dusty gray-blue can have the same hue and the same value but differ only in the amount of gray (chroma) in each color.

- Value

The difference between a light color (for example, light green) and a dark color (for example, dark green) of the same hue and chroma. This is the amount of pure white (tint) or pure black (shade) that has been added to the hue.

Words and Phrases

texture [ˈtekstʃə] *n*. 织物组织;织物质地,纹理,肌理

hand [hænd] *n*. 手感

construct [kənˈstrʌkt] *v*. 构造,结构

silhouette [ˌsiluːˈet] *n*. 侧面影象,轮廓

two-dimensional 二维的,平面的

clay [klei] *n*. 泥土,黏土

molding [ˈməuldiŋ] *n*. 模制,浇铸

potter [ˈpɔtə] *n*. 陶工,制陶工人

craft [kræft] *n*. 手艺,技巧,手工;手工艺

品行业
shiny [ˈʃaini] *adj*. 发光的；光亮的；有光泽的
snugly [ˈsnʌgli] *adv*. 舒适地，整洁干净地，紧密地
subtly [ˈsʌtli] *adv*. 柔和地，巧妙地，精细地
crepe [kreip] *n*. 绉绸；绉布
low-luster 光泽暗淡的，无光泽
camouflage [ˈkæməˌflaːʒ] *n*. 伪装，掩饰，保护色，迷彩
dye [dai] *n*. 颜料，染料；*v*. 染，染色
neutral [ˈnjuːtrəl] *n*. 中和色，不鲜明的颜色，与灰色相协调的颜色
dark color 深色，暗色
crisp [krisp] *adj*. 挺括，挺爽
appropriate [əˈprəʊpriət] *adj*. 适当的，适用的
perspiration [pəːspəˈreiʃən] *n*. 汗，汗水
spongy [ˈspʌndʒiː] *adj*. 海绵状的
lofty [ˈlɔfti] *adj*. 膨松的，弹性
thick [θik] *adj*. 厚，浓，深，密
outerwear [ˈautəweə] *n*. 外衣，外套，户外穿服装
intrinsic characteristic 内在特性
geometrics [ˌdʒiːəˈmetriks] *n*. 几何图形
flora [ˈflɔːrə] *n*. 花卉，植物
border [ˈbɔːdə] *n*. 边纹，镶边，滚边
even [ˈiːvən] *adj*. 平均的，均等的
light [lait] *n*. 淡色，浅色；光，光线；光泽
dark [daːk] *n*. 暗色，深色
spaced [speist] *adj*. 间隔的
clustered [ˈklʌstəd] *adj*. 成串的，成群的，成串的；聚集
muted [ˈmjuːtid] *adj*. 晕的；模糊的；暗淡的；柔和的
subdued [sʌbˈdjud] *adj*. 柔和的；缓和的
overall [ˈəuvərɔːl] *n*. 宽大罩衫，工作罩衣
motif [məʊˈtiːf] *n*. 基调，基本图案，基本色调
squint [skwint] *v*. 迷眼；斜眼
diagonal [daiˈægənəl] *adj*. 对角线的，斜的
hipline [ˈhiplain] *n*. 臀围，(女裙的)臀围部分
waistline [ˈweistˌlain] *n*. 腰节；腰围线，腰节线
chevron [ˈʃevrən] *n*. 山形，人字形；波浪形；V形
prints [prints] *n*. 印花，印花布
scale [skeil] *n*. 比例；缩尺；比例尺；等级；样卡
hue [hjuː] *n*. 颜色，色彩，色相
chroma [ˈkrəumə] *n*. 纯度；色品；色度
saturation [ˌsætʃəˈreiʃən] *n*. 色品度；纯度，纯色性；饱和度；浓度
dusty [ˈdʌstiː] *adj*. 不明朗的，灰暗的
gray-blue 灰蓝色
value [ˈvælju] *n*. 明度
tint [tint] *n*. 色，色彩，色泽，色度；淡色
shade [ʃeid] *n*. 色泽，色光，色调，色度，明暗程度

Notes

① Draping fabric on the dress form and sewing fabric to create garments teach a person how textiles react in various silhouettes. Converting textiles, flat two-dimensional material, into apparel provides essential training for the commercial designer exactly as working with clay and molding it into plates, bowls, and other items teaches the potter his craft.
在人台上立体裁剪制作服装，会帮助制作者了解纺织品不同造型后的效果。将平面

的纺织品转化成立体的服装，就如同陶工用泥土制陶盘、碗等陶器一样，是成衣设计师必不可少的训练。

② Shiny fabric reflects light and visually enlarges the area it covers, especially when a garment fits snugly. Fabric with a low luster reflects the light to a lesser degree than a shiny fabric and subtly enhances the wearer's face.

有光泽的面料反射光线，并在视觉上产生膨胀感，尤其是紧身服装。光泽暗淡的面料就没有那么强烈，能够柔和地烘托出穿着者的面部。

③ Crisp, absorbent fabrics, like cotton and linen, are appropriate for summer clothing because they absorb perspiration and allow the body to breath. Spongy fabrics with a soft and lofty hand feel warm because the thickness of the fabric retains the body's heat.

挺括的、吸湿性良好的面料因为吸汗和透气性适合制作夏装，例如棉和亚麻。柔软、膨松的海绵织物因为面料的厚度保留了人体的热量，从而给人温暖的感觉。

④ Pattern designs on fabric are created by color, lines, shapes, and spaces. They come in an endless variety-stripe, plaids, geometrics, floras, scenic, borders, and many others. The designs can be large or small, even or uneven, light or dark, spaced or clustered, muted or bold. All will affect how a fabric will look on you.

面料的图案由色彩、线条、形状和空间构成。图案种类非常丰富：条纹、格子、几何形、植物、风景、边纹以及其他多种类型。图案有大有小，有规则的或不规则的，有深有浅，有分散的也有集中的，有暗淡的也有醒目的，这一切都会影响服装着装后的效果。

⑤ When selecting striped or plaid fabrics, follow the basic theory of line and illusion. Identify the dominant stripe of the fabric. If you cannot identify it easily, try the squint test. As you squint at the fabric, note which stripe you see first. This will be the dominant one.

遵循基本的线条和视错原理来选择条纹或格子面料。如果你不能清楚地辨别主要的条纹，试着眯起眼睛观察，当你眯起眼睛去看时，最先跃入眼帘的就是主干条纹。

⑥ Select prints, stripes, and plaids that are in scale, or in proportional size, with your own body size. Small designs look best on the small to average person, but they look out of place and lost on a large figure. On the other hand, large designs are best worn by the average to tall person as these designs can overwhelm a small figure.

根据自己的体型选择大小适中或比例相称的印花、条纹和格子布。小花型适合小号体型和中等体型的人，大号体型的人穿着不合适。另一方面，大花型适合中等体型和高挑身材的人，小号体型会淹没在大花型中。

Discussion Questions

1. How can you successfully combine all the elements of design-line, shape, space, texture, and color-in clothes so they will look good on you?
2. What print themes are currently available in-store this season? Is there any correlation between the prints being used in men's wear, children's wear, and women's wear?

EXTENSIVE READING

FASHION DESIGN AND TEXTILES

The textile industries collaborate with a whole host of professionals, all experts within their own fields, from fashion to color and fibers. These experts play a pivotal role in the textile and fashion interrelationship: they guide the textile companies, designers and technologists, advising them about future ranges and predicting why or how their ranges will appeal to the consumer. Part of the mechanics of the relationship between fashion and textiles is rooted in trend and these influence the preliminary stages of fiber production (which generally operate two years ahead of a season).

Trends in fibers and fabrics develop from information gathered from professionals in fashion, textile mills, and other industry experts.

From this primary level, color and fabric trend information is then disseminated throughout the fashion and textile industries. Then initial judgements and choices in textiles focus on color. At this early stage views and considered opinions are drawn from industry experts such as the International Color Authority (ICA) and the Color Association of the United States. The fiber and fabric industries also resolve issues of texture, production and construction, which are typically informed by demands within fashion. The textile industry is steered by fabrication requirements that the fashion designer stipulates.

The fashion designer's relationship with fabric can be intensely personal. This intensity is very apparent at haute couture level, more so than at any other level, and is largely due to the fact that indulgence and personal expression can be afforded at this level. Designer can develop designs ideas by draping with a fashion fabric, the real material, rather than toiling with a fabric substitute like calico. In the late 1940s and early 1950s Jacques Fath①, Parisian couturier to royalty and movie stars, would create designs by moulding fabric directly onto a model, responding to the fabric as he worked, rather than sketching first.

Words and Phrases

collaborate with 合作
a whole host of 大量,众多
professional [prəˈfeʃənəl] *n.* 自由职业者,专业人士内行,专家
expert [ˈekspəːt] *n.* 专家,内行,能手
pivotal [ˈpivətl] *adj.* 关键性的
interrelationship [ˈintə(ː)riˈleiʃənʃip] *n.* 相互关系
consumer [kənˈsjuːmə] *n.* 消费者;用户
be rooted in 深植于
preliminary stage 初级阶段
disseminate [diˈsemineit] *v.* 撒播,传播,散布
initial [iˈniʃəl] *adj.* 最初的,初期的
steer [stiə] *v.* 引导,控制
fabrication [ˌfæbriˈkeiʃən] *n.* 加工,装配

结构物
stipulate [ˈstipjəˌleit] *v*. 确定，保证，规定；约定
intensely [inˈtensli] *adv*. 强烈地
intensity [inˈtensiti] *n*. 强烈
haute couture 高级女式时装
indulgence [inˈdʌldʒəns] *v*. 沉迷，沉溺
toile [twaːl] *n*. 试穿服装，样衣
substitute [ˈsʌbstitjuːt] *n*. 代用品；代替，替代；*adj*. 代用品的
calico [ˈkælikəu] *n*. 平布，白布，印花布，棉布
Parisian [pəˈrizjən] *adj*. 巴黎的，巴黎人的
couturier [kutyrˈje] *n*. 时髦女服商店，时髦女服商/女设计师
royalty [ˈrɔiəlti] *n*. 皇室，王族成员，特权阶层
model [ˈmɔdəl] *n*. 型，型号；时装模特，商品模特；款式，模型
sketch [sketʃ] *n*. 草图，速写

Notes

① Jacques Fath (1912～1954), French designer.
雅克·法特(1912～1954)，法国设计师。

LESSON 8 BASIC PRINCIPLES OF DESIGN / 设计原理

Proportion

Proportion is simply how the individual parts of a garment relate to the whole shape. The human body has many different contours, and a designer must modify parts of the garment to flatter the body. Sometimes this is achieved by emphasizing the natural body shape or sometimes by creating a new shape①. A designer will look at the space, dividing it by height and width to create a pleasing look. The classic natural waist proportion is a ratio of 3 (top) to 5 (skirt).

However, other proportion also works. The junior market in the mid 1990s featured a "body-doll" look for dresses. By making the bodice a shorter part of the design and the dress a longer part, the junior dress took on a younger, perhaps even a taller look. The empire waist exaggerates a young girl's proportion. Finally, broader shoulders creating a wedge proportion can help to play down a heavy set build②.

Balance

A designer divides a garment both horizontally and vertically. For the garment to appear appealing, the right amount of detail and emphasis must be distributed to each horizontal and vertical part. Too much or too little in one area of the garment makes it appear unbalanced.

Unity

Unity means that all elements included in the design work together and do not fight each other. For example, if a jacket has an off-center opening, the skirt underneath it should not have an on-center opening. Elements should look like that were planned, not a mistake.

Emphasis

Simply put, emphasis is the focal point, the center of interest of the garment, much as there is always a center of interest in a painting. It might be a fabric, a color, a detail, or trim, but it is the main reason your customers look at the garment!

Silhouette

Silhouette is the overall outside shape of a garment, and it is the most common element among all garments at a given point in time. When you see the padded shoulders and severe military suit silhouettes in a movie, you know that it was made in the 1940s.

The tight-waist full skirt, paired with a clinging knit top, is pure 1950s. Silhouette tends to change slowly, and it is one of the few features that will probably be alike among designers③. When the silhouette trend is oversized, for example, almost everyone will be doing oversized.

Line

The lines of a garment include the seams and edges of the garment that divide it. Lines usually create a kind of "visual illusion": longer, taller, or maybe slimmer. For example, princess seams create a slimming effect④. However sometimes, lines are used to create the illusion of weight. For example, a strapless look can be especially good for a thin woman because the shoulders appear wider. Strong asymmetrical lines tend to make a bold statement.

Words and Phrases

proportion [prəˈpɔːʃən] *n*. 比例，均衡，面积，部分

contour [ˈkɔnˌtuə] *n*. 轮廓，外形；轮廓线；略图

height [hait] *n*. 身高

waist [weist] *n*. 腰部

bodice [ˈbɔdis] *n*. 上衣片，大身，女装紧身上衣，紧身胸衣

shoulder [ˈʃəuldə] *n*. 肩宽，肩，肩胛骨

flatter [ˈflætə] *v*. (画像等的形象)美于(真人或实物)

emphasize [ˈemfəsaiz] *v*. 强调

emphasis [ˈemfəsis] *n*. 强调，重点

ratio [ˈreiʃiəu] *n*. 比；比例

junior [ˈdʒuːnjə] *n*. 少年，瘦小的女服尺寸，少女型

empire waist 帝国式腰线(高腰节)

exaggerate [igˈzædʒəreit] *v*. 夸大，夸张，使过大

wedge [wedʒ] *n*. 楔形

heavyset [ˈheviˈset] *adj*. 体格魁伟的

balance [ˈbæləns] *n*. 天平，秤，平衡

appealing [əˈpiːliŋ] *n*. 吸引力

unity [ˈjuːniti] *n*. 和谐，协调，统一

off-center 偏离中心的(地)

on-center 居中

opening [ˈəupəniŋ] *n.* 开襟，开门

focal point　焦点

padded shoulders　垫肩

military suit　军装

full skirt　宽裙，宽下摆裙，喇叭长裙，整圆裙

pair with　与……成对

clinging [ˈkliŋiŋ] *adj.* 紧身的，贴身的

knit top　针织上衣

oversize [ˈəuvəˈsaiz] *n.* 特大型，特大尺寸，尺寸过大

line [lain] *n.* 型，款式，纹路；线条，轮廓；系列；衬里

visual illusion　视错觉，错视

princess seams　公主线

strapless [ˈstræplis] *adj.* 无吊带，无肩带

asymmetrical [eiˈsimitrikəl] *adj.* 不对称的

bold [bəuld] *adj.* 鲜明的；清晰的；醒目的；大胆的，勇敢

Notes

① The human body has many different contours, and a designer must modify parts of the garment to flatter the body. Sometimes this is achieved by emphasizing the natural body shape or sometimes by creating a new shape.
人体由许多不同形状的部分组成，设计师必须调整服装的各个部位，通过强调人体的自然形状或创造出一个全新的造型，去美化人体体型。

② Finally, broader shoulders creating a wedge proportion can help to play down a heavy set build.
最后，相对较宽的肩膀形成的楔形则消除了造型的粗壮感。

③ The tight-waist full skirt, paired with a clinging knit top, is pure 1950s. Silhouette tends to change slowly, and it is one of the few features that will probably be alike among designers.
紧身针织上装和大喇叭裙的装束，是典型的 20 世纪 50 年代的造型。造型轮廓的变化是缓慢的，这也是不同设计师的作品容易雷同的少数原因之一。

④ The lines of a garment include the seams and edges of the garment that divide it. Lines usually create a kind of "visual illusion": longer, taller, or maybe slimmer. For example, princess seams create a slimming effect.
服装的线条包括服装的分割线和装饰线。线条通常会营造出一种"视错效果"：更长的、更高的或更苗条。例如，公主线会让人显得更加纤细。

Discussion Questions

1. Some people have a theory about the fashion trend. They call it "high tech, high touch". It means that when much of our world is highly technological, people need to surround themselves with things that feel good to touch and that remind them of the past. Do you agree with this theory? Why? Why not?

EXTENSIVE READING

FASHION TERMS

A Fad

A fad is a fashion that is very popular for a short time. Then suddenly it seems as if nobody is wearing it. A fad can be a color, such as mauve or chartreuse. It can be an accessory, such as earth shoes or rhinestone jewelry. Fads can also be an item of clothing. Short miniskirts and paratrooper pants with many pockets are examples of a fad. Fads can even be a certain look, such as "punk-rock" or "safari".

Style

Style refers to the shape of a particular item of clothing that makes it easy to recognize. Straight, A-line, and circular are all styles of skirts. Set-in, kimono and raglan are all styles of sleeves. Certain styles of garments are more fashionable at one time than another.

Classic

A classic is a traditional style that can stay in fashion for a very long time. The blazer jacket is a classic. Blue jeans are a classic. A tailored shirt and a cardigan sweater are classics.

A Status Symbol

A status symbol is an item of clothing that gives the weaver a special feeling of importance or wealth. Fashion designers are celebrities today. They put their names, initials, or symbols on the clothes they design to show that the clothes are special. People who wear those clothes are trying to communicate that they are special too. Some status symbols, such as a mink coat, are so expensive that very few people can afford them.

Old-Fashioned

Old-fashioned is a term that describes any style that we have grown tired of looking at. With today's instant communication, our brain receives a great deal of new information every day. Nothing seems new for very long. A garment can look old-fashioned to us in a very short time. However, the fashion pendulum swings back and forth so that some styles do return. Some examples of this are button-down shirts, skinny ties, and V-neck sweaters.

Words and phrases

fad [fæd] *n*. 流行一时的服装;流行快潮;一时风尚

popular [ˈpɔpjulə] *adj*. 流行的,普及的

mauve [məʊv] *adj*. 紫红色(的);淡紫色(的)

chartreuse [ʃɑːˈtrəːz] *adj*. 鲜嫩的黄绿色
earth shoe 大地鞋
rhinestone [ˈrainˌstəʊn] *n*. 人造钻石
jewelry [ˈdʒuːəlriː] *n*. 首饰,珠宝饰物
miniskirt [ˈminiˌskəːt] *n*. 超短裙,迷你裙
paratrooper pant 伞兵裤
punk-rock 朋克-摇滚
safari [səˈfaːriː] *n*. 瑟法里式,猎装,淡土黄色
A-line A字形线条,A型造型
set-in 另外缝上的,装袖
kimono [kəˈməʊnə] *n*. 连袖;连袖服装;和服;和服式女晨衣
raglan [ˈræglən] *n*. 插肩,插肩袖;套袖大衣,插肩袖大衣
jacket blazer 运动茄克
cardigan sweater 卡蒂冈毛衫,开襟式毛衫
tailored [ˈteiləd] *adj*. 定做的,合身的,精做的,精致的
celebrity [siˈlebriti] *n*. 名人,名流
initial [iˈniʃəl] *n*. 姓名的开头字母
symbol [ˈsimbəl] *n*. 象征,表征
mink coat 貂皮大衣
old-fashioned 过时的
pendulum [ˈpendjuləm] *n*. 钟摆
swing [swiŋ] *n*. 曲线形轮廓
back and forth 来回地
button-down shirt (前开襟纽扣的)传统衬衫
skinny [ˈskiniː] *adj*. 消瘦的;细窄的
V-neck V形领

LESSON 9 WHAT DO FASHION DESIGNERS DO / 时装设计师

Fashion designers usually specialize in menswear, women-wear or children-wear. Within each category they usually target a particular age range or customer lifestyle, such as menswear aimed at 25－40 years old. Some fashion designers are responsible for several different product types, particularly if they work in small companies or at a senior level in a large business. Designers can specialize in specific product type such as:

- Tailoring
- Casual-wear
- Sportswear
- Knitwear
- Jersey-wear
- Lingerie and underwear
- Swimwear
- Occasion-wear
- Club-wear
- Eveningwear
- Nightwear

Sometimes the job is narrowly focused on individual garment types, e.g. women's jackets in very large companies, such as suppliers to Marks&Spencer①, which sell styles in sizeable quantities②. It can be useful for the designer to gain detailed experience but some designers may feel restricted within a limited product area. In addition to garments, accessories are significant fashion items and designers are also required for millinery and footwear.

Fashion designers have several different responsibilities. In whichever area of the market they work, most fashion designers' jobs comprise the following tasks:

- Trend research
- Directional and comparative shopping
- Sourcing fabric and trims
- Design
- Range presentations
- Development meetings

Depending on the company they work for, fashion designers may also:

- Compile design specifications
- Cut patterns
- Fit garments
- Prepare garments for fashion shows

Design

When designing it is essential to take into account the lifestyle of the potential customer at whom the range is aimed. Fashion designers create ideas and make decisions in relation to:

- Silhouette
- Design detail
- Fabric
- Color
- Pattern
- Trims and fastenings

Design Development

Designers draw numerous initial design concepts before developing those with the most commercial potential. Variations on the most successful design ideas are sketched to view a series of potential alternatives[3]. Once a designer has chosen a suitable silhouette for a jacket, for example, different collar shapes, pockets and seams can be applied to the design before arriving at the definitive version. Fashion designers apply elements of the trends they have researched to products suitable for the target market. Designers may make their own decisions about which direction the designs should take or they may consult a senior designer to seek advice on the commerciality of their concepts. Whichever area of the market they cater for, fashion designers have to work within certain parameters when designing, such as:

- The customer's aesthetic tastes.

- The retailer's price range.
- The technical performance of the fabric and the completed garment.

The extent of the designer's creativity is constrained within these parameters to enable a commercial solution to a design brief to be found, whether the product forms part of a luxury designer collection or a value-led mass market range[④]. Other factors may be important in the design philosophy for certain companies and there is an increasing interest in ethical and environmental concerns within fashion and textiles, as the industry consumes enormous amounts of resources globally[⑤]. Appealing to contemporary consumer issues, some brands use organic fibers within their product ranges and a "green" stance can also be adopted by mass market companies within Fair Trade and Eco policies[⑥⑦].

Many designers try to stretch the boundaries in which their creative skills are confined, a quality which is often considered to go hand-in-hand with a creative personality. Fashion designers frequently receive criticism within the industry for lacking technical skills and concentrating too much on drawing. Designers who can consistently produce innovative concepts within technical and financial limits are viewed as valuable assets to the fashion industry.

图 20 A fashion designer specialize in casual trousers

Words and Phrases

specialize in 专门从事,专门研究
menswear ['menzweə] *n*. 男服
women-wear 妇女服装,女式服装
children-wear 童装
lifestyle ['laifˌstail] *n*. 生活方式
tailoring ['teilәriŋ] *n*. 缝制,裁剪业;裁缝工艺,成衣工艺
casual-wear 便服,轻便装
knitwear ['nitˌweə] *n*. 针织品,针织衣物
jersey-wear 针织物,针织坯布;弹力针织物
lingerie [ˌlænʒə'riː] *n*. 妇女贴身内衣
underwear ['ʌndəwɛə] *n*. 内衣,衬衣
occasion-wear 应时服装;特定场合服装
club-wear 俱乐部服装,娱乐服饰
evening-wear 晚礼服
nightwear ['naitweə] *n*. 晚服,夜间家常服
sizeable ['saizəbl] *adj*. 相当大的,可观的
restrict [ris'trikt] *v*. 限制;约束;
millinery ['miləˌneriː] *n*. 女帽,妇女头饰,女帽制造商,女帽商
footwear ['fʊtˌweə] *n*. 鞋袜(统称),鞋类(总称)
responsibility [riˌspɔnsə'biliti] *n*. 任务;职责
comprise [kəm'praiz] *v*. 包含,由……组成
compile [kəm'pail] *v*. 编制;搜集
take into account 把……考虑进去
potential [pə'tenʃəl] *adj*. 潜在的
fastening ['faːsniŋ] *n*. 扣合件
commercial [kə'məːʃəl] *adj*. 商业的
definitive [di'finitiv] *adj*. 确定的;明确的
cater for 投合,迎合

parameter [pə'ræmitə] *n*. 参数
taste [teist] *n*. 式样，风格，审美力，欣赏力
extent [iks'tent] *n*. 范围，程度
creativity [ˌkriːei'tivəti] *n*. 创造力，创造
constrain [kən'strein] *v*. 抑制，约束
value-led 价值引导的
mass market 大众市场；大规模市场
philosophy [fi'lɔsəfi] *n*. 基本原理；见解
ethical ['eθikəl] *adj*. 伦理的
environmental [enˌvaiərən'mentl] *adj*. 周围的，环境的
enormous amounts of 巨大的，庞大的
globally ['gləubəl] *adj*. 全球的
contemporary [kən'tempərəri] *adj*. 同时代的；当代的
organic fiber 有机纤维
stance [stæns] *n*. 姿态，态度，立场
boundary ['baundəri] *n*. 界线；边界；境界；范围
confine [kən'fain] *v*. 限制，局限
hand-in-hand 手牵手的，亲密的；并进的
frequently ['friːkwəntli] *adj*. 频繁的，常见的
criticism ['kritisizəm] *n*. 批评，非难；评论
concentrate ['kɔnsəntreit] *v*. 集中；使……集中于一点
consistently [kən'sistəntli] *adv*. 一贯地，一向，始终如一地
financial [fai'nænʃəl] *adj*. 财政的，财务的，金融的
asset ['æset] *n*. 有用的资源，宝贵的人[物]
fair trade 公平贸易，互惠贸易

Notes

① Marks & Spencer. M & S is a major British retailer, with over 885 stores in more than 40 territories around the world, over 600 domestic and 285 international.
玛莎百货是英国重要的跨国零售集团，全球 40 个地区拥有逾 885 家店铺，其中国内 600 多家，海外 285 家。

② Sometimes the job is narrowly focused on individual garment types, e.g. women's jackets in very large companies, such as suppliers to Marks & Spencer, which sell styles in sizeable quantities.
有时候，在大公司里，设计师仅需设计一类服装，如女茄克外套。例如玛莎百货的供应商，就只提供大量的一种款式的各种尺码（号型）。

③ Designers draw numerous initial design concepts before developing those with the most commercial potential. Variations on the most successful design ideas are sketched to view a series of potential alternatives.
时装设计师在计划商业生产之前就已经构思了大量的初步设计理念，最合适的设计思想将进一步深化并被绘制成一系列的草图，以观察可能出现的效果。

④ The extent of the designer's creativity is constrained within these parameters to enable a commercial solution to a design brief to be found, whether the product forms part of a luxury designer collection or a value-led mass market range.
设计师的创意程度受这些参数约束，这有助于发现商业解决设计主旨的方法：这个产品将成为奢侈品设计师发布会的一部分，还是价值引导的大众市场的成衣。

⑤ Other factors may be important in the design philosophy for certain companies and there is an increasing interest in ethical and environmental concerns within fashion and textiles, as the industry consumes enormous amounts of resources globally.

其他因素可能会在某些公司的设计理念中占有重要地位:服装业和纺织业所引发的道德和环保问题,消耗大量的地球资源等,均是日益关注的焦点。

⑥ Appealing to contemporary consumer issues, some brands use organic fibers within their product ranges and a "green" stance can also be adopted by mass market companies within Fair Trade and Eco policies.

为了迎合现代消费者的关注焦点,一些品牌在他们的产品中使用有机纤维,这种"绿色"的做法也被遵守公平贸易和生态政策的成衣公司采用。

⑦ Eco. Economic Cooperation Organization 经济合作组织

Discussion Questions

1. Have you ever wondered what inspired a fashion designer?
2. Have you ever wondered how many different fashion designers can come up with similar ideas at the same time?

EXTENSIVE READING

FASHION DESIGNERS AS ARTISTS

Considering the requirement that art is rare, one could argue that the couture designer is an artist. He or she creates a line from which individual copies may be ordered, much like the limited-edition print. Haute couture is exclusive, made of beautiful materials, and executed mostly by hand. While the commercial aspect is still present, the level of innovation is extremely high among the world's great couturiers, past and present. Madeleine Vionnet[①], for example, manipulated the qualities of supple fabrics, cutting on the bias to make extraordinarily beautiful clothes that look like liquid coverings for the body. Presently, art critics follow the creations of dozens of fashion designers for their artistic expressions of modernity. Some of these designers present their work as theatrical productions, akin to performance art. These costly, innovative fashion shows are part of the designer's vision for how to clothe the modern man or woman. Alexander McQueen[②]'s shows have been particularly notable in this regard.

Designers also appropriate images from well-known artworks for their fashion products, which seems to lend more credibility to their work. Piet Mondrian[③]'s color-block paintings, Pablo Picasso[④]'s cubist shapes, and Van Gogh[⑤]'s irises inspired Yves Saint-Laurent. Franco Moschino[⑥] used pop art images for his creations. Designers also collaborate with artists, and are inspired by artistic movements. Elsa Schiaparelli[⑦], for example, worked with Salvador Dali[⑧] and Jean Cocteau[⑨] to create Surrealist-inspired clothes.

Some fashion designers find expression in art media. Paul Poiret[⑩] painted after he no longer designed apparel. Ralph Rucci[11] experimented with watercolors and acrylics before deciding to become a fashion designer. Karl Lagerfeld[12] is a published photographer.

Words and phrases

limited-edition　限量版，限量发行的
exclusive [iks'klu:siv] *adj*. 唯一的，排他的
execute ['eksikju:t] *v*. 执行，实行，完成
innovation [ˌinəu'veiʃən] *n*. 改革，创新
extremely [iks'tri:mli] *adv*. 极端地，非常地
manipulate [mə'nipjuleit] *v*. 操纵，利用，操作，巧妙地处理
extraordinarily [iks'trɔ:dnrili] *adj*. 非凡，特别，非常，使人惊奇
art critic　艺术批评家
dozens of　许多的
modernity [mɔ'də:niti] *n*. 现代性，现代状态，现代东西
theatrical [θi:'ætrikəl] *adj*. 夸张的，戏剧性的
akin [ə'kin] *adj*. 同类的，同族的，同源的
innovative ['inəuveitiv] *adj*. 革新的，创新的，富有革新精神的
clothe [kləuð] *v*. 给……穿衣，盖上，赋予
notable ['nəutəbl] *n*. 著名人士
well-known　众所周知的，有名的
artwork ['a:twə:k] *n*. 插图，艺术作品
credibility [ˌkredə'biliti] *n*. 可信程度，确实性
color-block　色块
cubist shape　立体形象
iris ['aiəris] *n*. 鸢尾花
pop art　通俗艺术，波普艺术
Surrealist-inspired　超现实主义灵感来源的
acrylics [ə'kriliks] *n*. 丙烯酸树脂

Notes

① Madeleine Vionnet (1876～1975), French fashion designer, who during the 1920s and 1930s achieved critical acclaim when she developed the bias cut.
麦迪莲·维奥内(1876～1975)，法国时装设计师 20 世纪 20～30 年代因斜裁法而声誉鹊起。

② Alexander McQueen(1969～2010), is an English fashion designer.
亚历山大·麦柯奎恩(1969～2010)，英国时装设计师。

③ Piet Mondrian (1872～1944), Dutch painter, who carried abstraction to its furthest limits.
皮特·蒙德里安(1872～1944)，荷兰画家，他将抽象画发挥到极致。

④ Pablo Picasso (1881～1973), Spanish painter, who is widely acknowledged to be the most important artist of the 20th century.
巴勃罗·毕加索(1881～1973)，西班牙画家，作为 20 世纪最重要的艺术家广为人知。

⑤ Van Gogh (1853～1890), Dutch painter who exemplified the idea of artist as tortured genius.
梵高(1853～1890)，荷兰画家，痛苦的天才艺术家的典范。

⑥ Franco Moschino (1950～1995), Italian fashion designer. He became known for his irreverent clothing after opening his own business in 1983.
佛朗哥·莫斯奇诺(1950～1995),意大利时装设计师,1983 年开创自己的事业,服装因为反叛而闻名于世。

⑦ Elsa Schiaparelli (1890～1973), Italian fashion innovator and knitwear designer.
夏帕·瑞丽(1890～1973),意大利时尚创新者与针织服装设计师。

⑧ Salvador Dali (1904～1989), Spanish painter, writer, filmmaker, and designer, and one of the leading figures in the surrealist movement。
萨尔瓦朵·达利(1904～1989),西班牙画家,作家,电影人,设计师,超现实主义艺术运动的领军人物。

⑨ Jean Cocteau (1889～1963), French poet, novelist, dramatist, designer, and filmmaker, whose versatility, unconventionality, and enormous output brought him international acclaim.
让·谷克多(1889～1963),法国诗人,小说家,剧作家,设计师,电影制作人。多才多艺,独特的个性和大量的作品使他获得国际上的认可。

⑩ Paul Poiret (1879～1944), French fashion designer who was influential in the early part of the 20th century.
保罗·布瓦列特(1879～1944),20 世纪早期极具影响力的法国时装设计师。

⑪ Ralph Rucci (1957～), the first American designer to receive an invitation to the Paris haute couture shows since Mainbocher.
拉夫·鲁奇(1957～),自曼波彻以来,第一位应邀在法国开设时装展的美国设计师。

⑫ Karl Lagerfeld (1938～), German-born fashion designer, who has been a leading figure on the French fashion scene since the early 1970s.
卡尔·拉格菲尔德(1938～)出生于德国的时装设计师,自 20 世纪 70 年代始,成为法国时装界的领袖人物。

LESSON 10 THE EVOLUTION OF FASHION / 时装的演变

A new fashion is first worn by an innovator who wants to distinguish himself/herself from the accepted fashion norm. Usually, but not always, distinctive fashions are high-priced designer garments, often originating in Europe. The failure rate for experimental style is high. Some fashion starts "in the streets", created and worn by individualists who synthesize or design special outfits①.

Fashion professionals are constantly looking for new trends. As a new style begins "checking" or selling to trend setters, copying the style begins in earnest.

During the second phase, the trend setter stage, versions of the original style are made in different fabrics. Trend setters are often fashionable society personalities, movie stars, and other people who receive coverage in the press. Advertisements in glossy fashion magazines are common during this phase of the cycle.

The style next moves into the acceptance phase. The new trend is featured in window displays and image advertisements of better department and specialty stores. Buyers from volume retailers and manufacturers begin to copy the style at less expensive prices. At the peak of the acceptance phase, the style is widely copied, offered in many fabrics and colors, and often featured in high style mail order catalogs.

Classics evolve during the acceptance phase from basic garments which are worn by a wide variety of customers. Often, classics readily combine with other more novel items which greatly extend their acceptance. Classics are available at many price ranges, in a variety of fabrics, and have a wide appeal. The pleated trousers is an example of a classic which is modified slightly to reflect the style of a period②. For example, the trousers leg may be tapered or widened, cuffed or uncuffed, but the trousers itself has remained popular for many years.

The style enters the promotional phase as more and more copies (knock-offs) are made, often with the store's own label. Budget "resources" (manufacturers) cut the style in inexpensive fabrics to minimize costs③. Popularity decreases as less expensive fabrications are sold to mass merchandisers and discounters. The style is often found on sale racks and at jobbers.

The final demise of the fashion is the rejection phase of the fashion cycle. Now the style is so widely available in inexpensive copies that the mass of customers rejects it as an item too old to generate sales.

But waiting, almost every trend comes full cycle and is freshly interpreted by a new generation. Often fashion innovators create new looks by combining period pieces with a fresh eye and starting a trend anew several decades after it has saturated the market④.

Commercial designers analyze their customers, mentally placing them in a "niche", somewhere on the fashion curve. The designer looks at the fashion cycle as a whole, analyzing new trends and selecting the items, colors, and silhouettes that fit the lifestyle and fashion cycle of the customer.

The Fashion Cycle

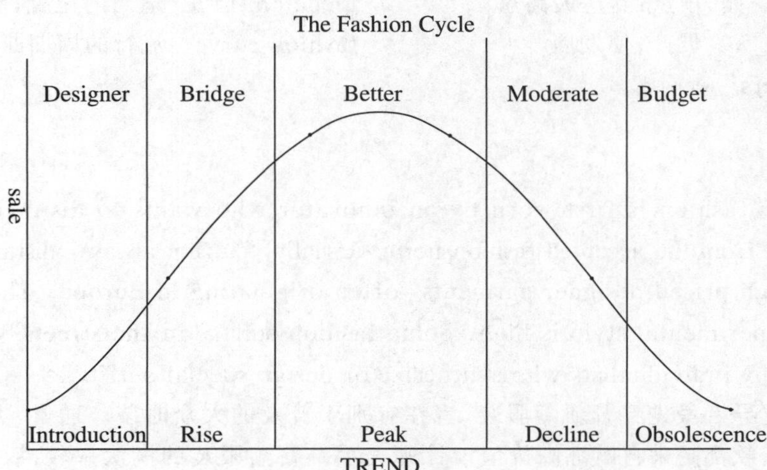

Words and Phrases

innovator [ˈinəveitə] *n*. 改革者,革新者
distinctive [diˈstiŋktiv] *adj*. 明显不同的,特别的,突出的
high-priced 高档价格
the failure rate 失败率
experimental [iksˌperiˈmentəl] *adj*. 实验的;经验的
individualist [ˌindiˈvidjuəlist] *n*. 个人
synthesize [ˈsinθiˌsaiz] *v*. 综合;合成
outfit [ˈautfit] *n*. 服装,全套服装
trend setter (服装式样)创新人
in earnest 认真地,诚挚地
personalities [ˌpəːsəˈnælitiz] *n*. 知名人士,名流
coverage [ˈkʌvəridʒ] *n*. 新闻报导
press [pres] *n*. 出版社,出版机构;记者;新闻舆论
glossy [ˈglɔːsiː, ˈglɔsi] *adj*. 有光泽的,光鲜的
feature in 占重要位置
window display 橱窗展示,橱窗陈列
department [diˈpaːtmənt] *n*. 部门
specialty store 专卖店
volume [ˈvɔljuːm] *adj*. 大量的,大批量
retailer [riˈteilə] *n*. 零售商
peak [piːk] *n*. 巅,顶点;到达最高点
evolve [iˈvɔlv] *v*. 展开;使发展
pleated trousers 有褶裤
modify [ˈmɔdifai] *v*. 修正,修改
slightly [ˈslaitli] *adv*. 些微地,苗条地,瘦长地
taper [ˈteipə] *adj*. 锥形的;使成锥形
cuffed [kʌfd] *adj*. 翻边的,折边的
promotional [prəˈməuʃənl] *adj*. 促销的,推销的
budget [ˈbʌdʒit] *v*. 预算,安排
knock off 翻制设计,翻印本
minimize [ˈminimaiz] *v*. 将……减到最少,最小化
popularity [ˌpɔpjuˈlæritiː] *n*. 流行,普及
fabrication [ˌfæbriˈkeiʃən] *n*. 制造,生产,制作,加工;成品
mass merchandiser 超型市场
discounter [ˈdiskauntə] *n*. 折扣商店
sale rack 展销,促销
jobber [ˈdʒɔbə] *n*. 批发,批发商
demise [diˈmaiz] *n*. 结束,完结
rejection [riˈdʒekʃən] *n*. 抛弃
fashion cycle 时装界
available [əˈveiləbl] *adj*. 可利用的,可获得的,有效的
saturate [ˈsætʃəreit] *v*. 渗透,浸透,充满,使饱和
niche [nitʃ] *n*. 适当的位置,恰当的处所
fashion curve 流行的周期曲线

Notes

① A new fashion is first worn by an innovator who wants to distinguish himself/herself from the accepted fashion norm. Usually, but not always, distinctive fashions are high-priced designer garments, often originating in Europe. The failure rate for experimental style is high. Some fashion starts "in the streets", created and worn by individualists who synthesize or design special outfits.
标新立异者最先穿着新潮服装,旨在有别于普及的大众时装。通常,大多数别致的服装新款都是来自欧洲高价位设计师之手。新潮服装的淘汰率较高。有些新潮服

装是来源于“街头服饰”，由极具个性的时尚达人组合、设计、搭配而成。

② Classics evolve during the acceptance phase from basic garments which are worn by a wide variety of customers. Often, classics readily combine with other more novel items which greatly extend their acceptance. Classics are available at many price ranges, in a variety of fabrics, and have a wide appeal. The pleated trouser is an example of a classic which is modified slightly to reflect the style of a period.

流行过程中，当基本款被大多数人接受时形成经典服装。通常，经典服装与其他新鲜元素组合可以让他们重新焕发生机。经典的服装可以出现在任何价位、任何面料，可以有很多种形态。褶裤就是这样的一个例子，稍微调整一下，就又能表现一个时期的流行，由各种各样的面料制作，以更广的价格范围赢得大众的需求。

③ The style enters the promotional phase as more and more copies (knock-offs) are made, often with the store's own label. Budget "resources" (manufacturers) cut the style in inexpensive fabrics to minimize costs.

当出现越来越多商家自制的复制品时，意味着这个款式已经进入了推广阶段。制造商的预算部门会策划使用廉价面料来生产，以降低成本。

④ Often fashion innovators create new looks by combining period pieces with a fresh eye and starting a trend anew several decades after it has saturated the market.

时装创新者常常从全新的视角观察，将某个时期的流行款式重新组合而形成新的风格，等市场上这种款式已经饱和时，再引导另一个几十年的流行。

Discussion Questions

1. What influences the fashion?
2. What fads are currently popular in your school?
3. What criteria do people use to decide when clothing is old-fashioned or no longer in style?

EXTENSIVE READING

FORECASTING IN THE APPAREL INDUSTRIES

Fashion and change are synonymous, and no publication on fashion goes to press without announcing some new trends. The forecaster stands in the middle of a constantly shifting fashion scene and translates ambiguous and conflicting signals to provide support for business decisions. Professional forecasters are employed by advertising agencies, consumer-product companies, trade groups, corporations along the textile/apparel pipeline, forecasting services, and as consultants. Forecasters work at all stages of the textile/apparel supply chain on timelines that vary from a few months in advance of the sales season to ten years ahead of it.

In companies today, forecasting must be a team effort, with information shared between functional groups including design, merchandising, marketing, sales, and promotion, so that the right product gets produced and distributed at the right time to a target consumer. In the world of fashion, improving the success rate of new merchandise, line extensions, and retailing concepts by only a few percentage points more than justifies the investment of time and money in forecasting. Today's executive must be skilled in the use of an array of quantitative and qualitative forecasting methods that support the decision-making process.

Fashion forecasters share common ground: they believe that by keeping up with the media, analyzing shifts in the culture, interviewing consumers, and dissecting fashion change, they can spot trends before those trends take hold in the marketplace. By anticipating these changes, forecasters allow companies to position their products and fine-tune their marketing to take advantage of new opportunities. Major companies are becoming more and more dependent on this kind forecasting because traditional forms of purely quantitative forecasting are less applicable to an increasingly volatile and fragmented marketplace. Two factors make forecasting more important today than ever before: the global nature of apparel production and marketing, and the shift to time-based strategies of competition. Forecasting creates competitive advantage by anticipating trends, aligning product development with consumer preferences, and facilitating the timely arrival of products in the marketplace.

Words and Phrases

synonymous [si'nɔnəməs] *adj*. 同义的
publication [ˌpʌbli'keiʃən] *n*. 出版物；出版
announce [ə'nauns] *v*. 发布，宣布，预告
forecaster ['fɔːkaːstə] *n*. 预测者
shift [ʃift] *v*. 变化，更替，转移，改变 *n*. 开关；变化
ambiguous [æm'bigjuəs] *adj*. 不明确的，模棱两可的，模糊的
conflict ['kɔnflikt] *v*. 战斗，冲突，矛盾，抵触
signal ['signəl] *n*. 信号，暗号，标志
advertising agency 广告公司，广告社
consumer-product company 消费产品公司
corporation [ˌkɔːpə'reiʃən] *n*. 公司，企业
pipeline ['paipˌlain] *n*. 流水线，商品供应线
consultant [kən'sʌltənt] *n*. 顾问
supply chain 供应链
timeline 时间轴
extension [iks'tenʃən] *v*. 延长，扩充，范围，扩展名
percentage [pə'sentidʒ] *n*. 百分比，比率，部分，可能性
justify ['dʒʌstifai] *v*. 证明……有道理，为……辩护
executive [ig'zekjutiv] *n*. 执行部门，执行者，经理主管人员 *adj*. 行政的
quantitative ['kwɔntitətiv] *adj*. 数量的，定量的
qualitative ['kwɔlitətiv] *adj*. 性质的，质的，定性的
decision-making 决策，决策的
analyze ['ænəlaiz] *v*. 分析，细察，分解
interview ['intəvjuː] *n*. 面谈，访问，接见，面试
dissect [di'sekt] *v*. 解剖，切开

spot *vt*. 认出,发现
anticipate [æn'tisipeit] *v*. 预期,希望,预料
fine-tune 微调
take advantage of 利用
opportunity [ˌɔpə'tjuːniti] *n*. 机会,时机
applicable ['æplikəbl] *adj*. 合适的,可应用的
increasingly [in'kriːsiŋli] *adv*. 逐渐地,渐增地
volatile ['vɔlətail] *adj*. 可变的,不稳定的
fragment ['frægmənt] *n*. 碎片,片段
time-base 时间基线;时间坐标
strategy ['strætidʒi] *n*. 战略,策略
preference ['prefərəns] *n*. 偏爱,喜爱
facilitate [fə'siliteit] *v*. 使容易,促进,帮助

LESSON 11 TOOLS TO PREDICT FASHION / 流行预测机构

Collection Reports

Collection reports are the most immediately available source of information on designer collections, but they can be expensive. These reports① provide detailed sketches of new designs, including color and fabric selections, and all of them are based on a recent designer fashion show. Many companies will have these reports out two weeks after a designer's show.

图 22 Pantone color system

Trend Reports

Several companies compile what are called trend reports; these sources include the Tobe Report① and Block Note. One other source, Mode, provides a website that has information on all of the available trend reports. Trend reports are similar to collection reports in that they provide sketches, pictures, and fabric/color swatches to assist the fashion buyer and designer in decision making②. Trend reports are written by independent companies and are not based on any one designer's collection.

Similar to trend reports, color services provide predictive information on popular new colors to be used for a coming season. Companies that specialize in color forecasting include the International Color Authority. The Color Box, and Concepts in Color. Color systems are used in color prediction. Two color systems are those provided by Pantone and SCOTDIC(Standard Color of Textile Dictionaire Internationale de la Couleur). Both of these companies keep records in the form of dyed fabric of all colors used in the past.

Websites

There is a difference between a fashion forecasting website and one that provides simple trend information to designers and buyers. Fashion forecasting websites typically require fee-based subscriptions for access. These types of websites are geared toward buyers, which is why they charge a fee. There are also many commercial websites more geared to the costumer, but buyers will view these too.

The information provided by these available online resources, be they are fee-based or free, will help fashionist as design better lines or purchase the right merchandise for their stores③.

Trade Publications

Trade publications can be newspapers, magazines, or journals that are written specifically with the fashion professional in mind.

It is extremely important for people in the fashion industry to read such trade publications to learn about new trends, mergers, and other changes within the industry. Also important to read a fashion magazines, sometimes called editorial publications or consumer publications.

In addition to trade publications, trade associations may provide information on trends as well. A trade association is a professional organization of businesses or manufacturers that formed to promote their business or industry or to adopt uniform standards.

Movies, Music, and Television

Most good trend forecasters will be aware of styles that movie and music stars are wearing. Besides providing inspiration for designers, what stars wear is useful in forecasting trends. Often, celebrities bring fashion to the mainstream, to the forefront for the mass-marketing of garments, rather than starting the actual trend④.

Street Fashion

Finally, street fashion, or what everyday people are wearing, can be an indicator for trend forecasters. Simple observation techniques such as watching people in malls, at concerts, and at other events can discover new fashion trends.

图 23 Fashion shows can start fashion trends and make a current "look" seem old-fashioned

Words and Phrases

compile [kəm'pail] *vt*. 编辑,搜集(资料)
mode [məud] *n*. (服式的)式样,风尚,风气,流行
fee-based 以收费为基础
subscription [səb'skripʃən] *n*. 订阅
gear toward 专门朝向某事
gear to 适合于
mainstream ['meinˌstriːm] *n*. 主流
forefront ['fɔːˌfrʌnt] *n*. 最前部,最前沿
street fashion 街头服装,街头服饰
observation [ˌɔbzəː'veiʃən] *n*. 注意,观察
fashionist ['fæʃənist] *n*. 时尚追随者;时装店主

Notes

① The Tobe Report, first published in 1927, is one of the world's most prestigious fashion merchandising consulting companies. An international weekly publication, the Tobe Report covers women, men, and children's ready-to-wear and accessories. It forecasts trends through analysis of consumer behavior and retail intelligence.
Tobe Report 于 1927 年首刊,世界著名时装业咨询机构之一。国际性周刊,内容覆盖女装、男装、童装及配饰品。通过分析消费者行为和零售情报,预测流行趋势。

② Several companies compile what are called trend reports; these sources include the Tobe Report and Block Note. One other source, Mode, provides a website that has information on all of the available trend reports. Trend reports are similar to collection reports in that they provide sketches, pictures, and fabric/color swatches to assist the fashion buyer and designer in decision making.
有几家机构编写流行趋势:Tobe Report 和 Block Note,此外,Mode 这家公司提供的网页也包含所有可用的流行趋势。流行趋势报告类似于时装发布,预测机构为时装买手和设计师提供各类效果图、照片和面料/色彩小样,以协助商业决策。

③ The information provided by these available online resources, be they fee-based or free, will help fashionist as design better lines or purchase the right merchandise for their stores.
这些可用的网络资源提供的信息,无论收费与否,都能帮助时装从业者们更好地策划设计生产,或经营合适的产品。

④ Most good trend forecasters will be aware of styles that movie and music stars are wearing. Besides providing inspiration for designers, what stars wear is useful in forecasting trends. Often, celebrities bring fashion to the mainstream, to the forefront for the mass-marketing of garments, rather than starting the actual trend.
许多优秀的预测者会关注电影电视明星的衣着,除了给设计师带来灵感,明星们的穿着还有助于预测流行趋势。名流常常将顶尖时尚展示给大众和高级成衣市场,但这并不是真正流行趋势的开始。

Discussion Questions

1. How do you decide when something is old-fashioned?
2. What is fashion?

EXTENSIVE READING

TRADES SHOWS

There are a number of specialty forcasting services available to merchandisers on a subscription basis. These subscriptive services provide periodic reporting on yarn, fabric, color, silhouette, and retail sales trends. A sampling of these services follows:

- Hong Kong Fashion Week, Hong Kong Convention & Exhibition Center, Wanchai, Hong Kong. Organizer: Hong Kong Trade Development Council, 38th Floor, Office Tower, Convention Plaza, 1 Harbour Road, Wanchai, Hong Kong.
- Bangkok International Fashion Fair, Bangkok International Trade & Exhibition Center, Bangkok, Thailand. Organizer: Department of Export Promotion, Ministry of Commerce, Royal Thai Government, 22/77 Rachadapisek Road, Chatuchak, Bangkok 10900, Thailand.
- International Kids Fashion Show, Jacob K. Javits Convention Center, New York City. Organizer: The Larkin Group, 485 Seventh Avenue, Suite 1400, New York, NY 10018.
- Prêt a Porter, Paris, Porte de Versailles, Hall 7. Organizer: SODES, 5 rue de Caumartin, 75009 Paris.
- Herren Mode/Inter-Jeans, Koln Messe, Messeplatz 1, 50679 Cologne, Germany. Organizer: Koln Messe, Messe und Austellungs Ges. mb. h., Koln, Messerplatz 1, D-50679 Koln, Postfach 210760D-50523 Koln, Germany.
- CPD Dusseldorf, Dusseldorf, Germany. Organizer: Igedo Company, Stockumer Kirchstrasse 61, D-40474 Dusseldorf, Germany.
- ISPO, New Munich Trade Fair Center, Germany. Organizer: Messe Munchen GmbH, Messeqelande, D-81823, Munchen, Germany.
- Pitti Filatti (fabrics), Fierra Milano, Italy. Organizer: Pitti Immagine, Via Faenza, 109, 50123 Florence, Italy.
- MAGIC West, Las Vegas Convention Center, Las Vegas, NY;
- WWDMagic, Sands Expo Center, Las Vegas, NV. Organizer: MAGIC International, 6200 Canoga Avenue, Suite 303, Woodland Hills, CA 91367.

图 24 Trade exhibition is useful to the fashion business

- UK Fashion Fair, London, England. Organizer: Igedo Company, Stockumer Kirchstrasse 61, D-40474 Dusseldorf, Germany.
- Intermezzo Collections, Piers 90 and 92 at the Show Piers on Hudson, 12th Avenue and 55th Street, New York City. Organizer: ENk International, 3 East 54th Street, New York, NY 10022.

Words and Phrase

periodic [ˌpiəriˈɔdik] *adj*. 周期(性)的,定期的,循环的

LESSON 12 THE FASHION-DESIGN PROCESS / 时装设计过程

The fashion-design process is a lengthy one and can have many variations, depending on the type of company. The general steps in design include:

1. Research and planning of the line.
2. Creating the design concept.
3. Development of the designs.
4. Production planning and production.
5. Distribution of the line.

Research and Planning of the Line

This is the stage during which designer research fabric trends, color trends, past sales, and goals of the organization. Some companies will produce their own merchandise, called private label; others may only buy designs from manufacturers. In-house design, or private-label design, occurs when retailers have their own designers create clothing for them. Designers will review fashion magazines, look at color forecasts, and even watch celebrities to see what the new hot fashion will be. The first step sets the groundwork for the rest of the process.

Creating the Design Concept

The next step in the design phase is to create the concept. For the most part, the concept will actually be sold to in-house buyers before production even begins. In this phase, the designer will develop sketches, gather fabric samples, and develop a line, usually eight pieces, that go together. A line board is then presented to decision makers in the organization, who then must approve it before the concept enters into the next steps of design①.

Much of the industry is moving toward the use of computer aided design in the development of line boards. CAD can also assist in the making and cutting of patterns. Fashion manufacturers are also experimenting with the use of CAD for everything from design to production to shopping and distribution management of the products[②].

Development of the Designs

After the approval to go ahead with the designs has been received, the designer or patternmaker will produce a pattern for that garment. There are two ways to make a pattern. The first way is with a flat pattern, in which the design is made out of hand-cut pieces of heavy paper. Draping, the second way, consists of actually arranging the fabric on a person or on a body form and pinning right on the form to make the pattern.

After the pattern has been made, a sample is generally made out of muslin, an inexpensive fabric to test for fit. Then, if the designer is happy with the fit, a sample sewer will then make the garment with the actual fabric. The sample sewers are often in-house, or sometimes they can work for an overseas manufacturer. During this phase, the price of the product will also be determined.

Production Planning and Production

Due to the expansion of the global economy, most manufacturing is done overseas. Because of this, the designer will develop a specifications package, sometimes called a spec package[③]. A spec package includes drawings and specific information related to the details of the garment. The details can include, for example, the type of closures used or the type of collar. Another element of the spec package is specific information as to how the garment should be put together, meaning, where the seams should actually be. Sometimes, the manufacturer will send a sample of the garment made from the spec package, to insure proper fit and expectations. An important part of this process includes the ability to communicate effectively with overseas producers. Once the spec package is sent, the manufacturer will cut, sew, and finish the product. It will then be shipped as requested.

Distribution of the Line

Distribution of the line can also be considered the marketing or actual selling of the line. Distribution is defined as the manner in which goods actually get to the customer. Distribution includes the process of working with overseas vendors, shipping the goods from overseas, and actually placing the goods in stores.

Words and Phrases

private label　商店标签;独立设计师标签
muslin [ˈmʌzlin] *n.* 平纹细布;[美]棉布,薄纱织物
in-house　公司内部的,内部的,室内的
groundwork [ˈgraundwəːk] *n.* 基础
board [bɔːd] *n.* 熨衣垫板
flat pattern　平面纸样
hand-cut　手工裁剪
closure [ˈkləʊʒə] *n.* 服装闭合件
manner [ˈmænə] *n.* 方式,方法;态度

vendor [ˈvendə] *n*. 厂商,卖家

Notes

① The next step in the design phase is to create the concept. For the most part, the concept will actually be sold to in-house buyers before production even begins. In this phase, the designer will develop sketches, gather fabric samples, and develop a line, usually eight pieces, that go together. A line board is then presented to decision makers in the organization, who then must approve it before the concept enters into the next steps of design.
接下来的设计阶段即是形成设计理念,在生产之前,这个理念实际上会卖给公司内部的买家。这个阶段,设计师将要绘制效果图,收集面料样本,设计一个系列,通常一共八套款式。这个系列策划然后被送至公司内部的决策者,并决定是否进入下一个阶段。

② Much of the industry is moving toward the use of computer aided design in the development of line boards. CAD can also assist in the making and cutting of patterns. Fashion manufacturers are also experimenting with the use of CAD for everything from design to production to shopping and distribution management of the products.
许多工厂在系列的策划中采用计算机辅助系统,CAD 系统也支持纸样的绘制和裁剪。服装制造商也具备了采用 CAD 进行产品设计、生产、船运和销售管理的经验。

③ Due to the expansion of the global economy, most manufacturing is done overseas. Because of this, the designer will develop a specifications package, sometimes called a spec package.
由于全球经济化的扩展,许多加工是在海外完成的,基于此,设计师需要制作工艺规格文件包,有时简称工艺单袋。

Discussion Questions

1. Discuss the general steps in the fashion-design process.

EXTENSIVE READING

SKETCH

Many careers in fashion require drawing skills. Designers sketch their ideas so they can be interpreted by pattern makers. The working sketch, also known as a "flat", must accurately show the silhouette and style details. The sketch should be drawn to scale so technicians can accurately gauge the size and placement of the details. Designers often write specifications alongside the sketch to describe the garments in greater detail.

Sample makers use the working sketch as a guide when constructing the garment. The working sketch is used to identify patterns. Pattern makers or assistant designers sketch the garment again for cost sheets, catalogs, sales sheets, and the like.

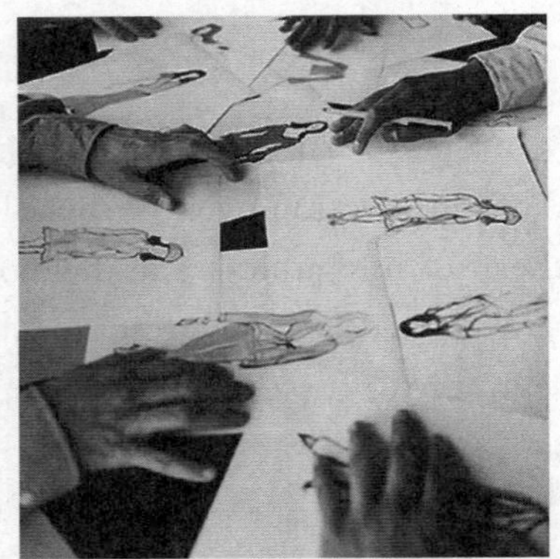
图 25 The company selects the right sketch which expresses their idea and concepts

Buyers who can sketch will often make picture notes of garments they are purchasing. The trend in retailing is to create merchandise for a store. Buyers copy styles from major designer lines which have sold well in their stores and have the garments manufactured in quantity for a lower price than the original. The copied garment is called a knock-off.

Fashion consultants are specialists in selecting garments which flatter persons who have special wardrobe requirements. They often work with custom designers and must create working sketches to advise their clients on what to wear and guide the dressmaker who sews the garment.

Fashion illustrators have a different purpose in drawing a garment. They want to enhance the original design concept. Illustrators emphasize current ideals of beauty to create a stylized, exaggerated version of reality which is designed to sell the garment. Fashion illustration requires a high degree of artistic achievement. Rendering the fabric, selecting the appropriate model pose, makeup, hair style, and attitude require a great deal of drawing ability.

Words and Phrases

flat [flæt] *n.* 平面图，效果图
sample maker 样衣制作师
pattern maker 样板师，纸板师，打样师
scale [skeil] *n.* 比例；缩尺；比例尺；等级；样卡
specification [ˌspesifiˈkeiʃən] *n.* 规格，规范；明细表；(产品)说明书
assistant designer 助理设计师
cost sheets 成本核算表
sales sheets 销售(报)表
fashion consultant 时装顾问，服饰顾问
wardrobe [ˈwɔːdrəub] *n.* (个人的)全部服装，行头；衣柜，衣橱
custom designer 定制服装设计师
dressmaker [ˈdresˌmeikə] *n.* 裁缝工，服装工；女服裁缝师
fashion illustrator 时装画
stylized sketch 效果图
artistic achievement 艺术成就
render [ˈrendə] *v.* 表现，反映
pose [pəuz] *n.* 造型，姿态
makeup [ˈmeikˌʌp] *n.* 化妆；化妆品
hair style 发式，发型

LESSON 13 MODELS / 模特

The model is an indispensable part of fashion; without the model, clothes could not be fitted, tried out, appraised and shown to buyers. The public only sees the supermodels at spectacular events in the fashion capitals of the world, but there are models at all levels. Although there are both male and female models, it is really the model girl who is most apparent as an adjunct to the fashion process.

The Model Girl

Despite talk of anorexia and exploitation of the young, the slim, perfectly proportioned figure of a model is of paramount importance. Not only does she have to be slim and tall but she must also have good shoulders, a small waist, slim hips and extra long legs: a glance at any fashion illustration over the past 60 years will tell you why-these attributes flatter the clothes① . What is not realized is that if a garment is fitted on a model with these characteristics, the garment will take on this shape and impart it to the less well-proportioned wearer.

图 26 Supermodels in a Spring/Summer fashion show in Paris

Models can be showing or photographic; often a good showing model is not photogenic and a photographic girl may not be tall enough for showing.

Models at all levels are needed, and to this purpose there are model agencies in most cities. In the past, couture houses would employ house models, one or two on a permanent basis to do all the fittings and give small in-house showings. They would also hire several other model girls for the season. This, then, was the method for a girl to gain experience.

The top agents such as Models One in London or Elite in New York deal only with those girls who are likely to show on the international catwalks or be photographed for top fashion magazine or large advertising campaigns② . Many models make their money from the much more mundane catalogue work, which nevertheless pays well③ . Other levels of model are used for trade fairs such as car shows, boat fairs and the many fashion fairs, for example The Clothes Show Live in Birmingham, UK.

The Male Model

In many ways the male model, from a fashion point of view, is less important than his female counterpart. It is quite possible for catwalk shows to take a man off the street, as long as he is tall enough, fairly good looking and can walk in a straight line④ . For photographic work, however, it is another story; the male model has to be extremely good-looking, photogenic, and have that extra something that can sell a product. In this area men can reach supermodel status, especially in huge advertising campaigns, as Werner Schreyer did with Hugo Boss in 1995.

Words and Phrases

indispensable [ˌindisˈpensəbl] *adj.* 不可缺少,绝对必要的

try out 试验，试穿
appraise [ə'preiz] *v*. 评价，估价
supermodel ['sjuːpəˌmɔdəl] *n*. 超级模特
spectacular [spek'tækjulə] *n*. 公开展示的，展览物
adjunct ['ædʒʌŋkt] *n*. 附属物，附件，助手
anorexia [ˌænə'reksiːə] *n*. 厌食症
exploitation [ˌeksplɔi'teiʃən] *n*. 开发，开采
paramount ['pærəˌmaʊnt] *adj*. 最重要的，最高的
attribute [ə'tribjuːt] *v*. 认为……属于
impart [im'paːt] *v*. 给予，传授，告知
catwalk ['kætwɔːk] *n*. (舞台等)天桥
photogenic [ˌfəutəu'dʒenik] *adj*. 适宜于拍照的，拍照效果好的，特别上镜的
mode agency 模特介绍所，模特经纪所
campaign [kæm'pein] *n*. 竞选运动
mundane [mʌn'dein] *adj*. 现世的，世俗的
nevertheless [ˌnevəðə'les] *adv*. 然而，虽然如此
car shows 车展
fair [fɛə] *n*. 博览会，商品展览会，商品交易会
counterpart ['kauntəpaːt] *n*. 对手

Notes

① Not only does she have to be slim and tall but she must also have good shoulders, a small waist, slim hips and extra long legs: a glance at any fashion illustration over the past 60 years will tell you why-these attributes flatter the clothes.
她不仅要苗条和高挑，而且还必须拥有完美的肩形、纤细的腰围、窄小的臀部和修长的腿型。回顾过去 60 年的时装图例，你会发现这些体形特点都是为了让服装看上去更好。

② The top agents such as Models One in London or Elite in New York deal only with those girls who are likely to show on the international catwalks or be photographed for top fashion magazine or large advertising campaigns.
顶级的模特经纪所，例如纽约的 Models One 或 Elite，仅仅从事那些可能会出现在国际天台上或顶级时装杂志或大型广告活动中的女模特们的经纪事务。

③ Many models make their money from the much more mundane catalogue work, which nevertheless pays well.
许多模特从事更加平民化的待售品目录的工作，以获取高额报酬。

④ In many ways the male model, from a fashion point of view, is less important than his female counterpart. It is quite possible for catwalk shows to take a man off the street, as long as he is tall enough, fairly good looking and can walk in a straight line.
从时装的视角来看，男模在许多方面没有女模重要。只要他足够高，相貌英俊，可以走一条直线，极有可能就会在街头被发现，登上天台走猫步去了。

Discussion Questions

1. List the famous model agents in your city.

EXTENSIVE READING

FASHION SHOWS

Formal fashion shows are used to introduce new products for each season. Fashion shows by designers such as Armani① or Dior② are a mainstay of the industry. At a formal fashion show, the audience includes members of the press, buyers, and private customers. Apparel manufacturers also hold fashion shows, but typically without the glitz of a couture fashion show, such as one for Dior. The attendees of these shows tend to be buyers, thus such as an event provides an opportunity to gain fruitful sales for the season. Formal fashion shows are generally expensive to produce, which is why many smaller designers choose not to stage them.

图 27 Fashion shows can start fashion trends and make a current "look" seem old-fashioned

Informal fashion shows can be inexpensive to produce. They can include those that take place in a mall or inside a department store. Or a fashion show might include several designers, who pay a fee to have their garments shown. An example of this kind of show is the Erotic Fashion Show, produced in Seattle. It features designers of leather and other fetish and gothic clothing. For some shows like this, even the attendees must pay an entry fee, which generates income for its producers.

A trunk show is when merchandise is brought into a store and featured for a few days. Often, customers will get special invitations to attend these events, and sometimes the designers will be on hand to answer questions. This is an inexpensive way to gain publicity and is typically used in conjunction with direct marketing to inform the customers about the show.

Words and Phrases

mainstay [ˈmeinˌstei] *n.* 支柱

glitz [glits] *adj.* 闪光的，炫目

attendee [ˌætenˈdiː] *n.* 与会者

fruitful sales 富有成果的销售

Erotic Fashion Show 艳情时装表演

Seattle [siˈætl] *n.* 西雅图

fetish [ˈfetiʃ] *n.* 偶像

gothic clothing 哥特式服装

trunk show 展销

in conjunction with 连合，结合，联系

Notes

① Giorgio Armani (1934～), Italian fashion designer whose clothes combine understated elegance, quality tailoring, and practicality.

乔治·阿玛尼，意大利时装设计师，其作品将简约优雅、精致和实用融为一体。

② Christian Dior (1905～1957), French couturier, born in Granville, and educated for the diplomatic service at the École des Sciences Politiques in Paris.
克里斯汀·迪奥，法国高级时装设计师，出生于格朗维尔，曾就读于巴黎科学政治学院，攻读外交官事务专业。

LESSON 14 ACCESSORY DESIGN / 配饰设计

Fashion accessories are decorative items that supplement one's garment, such as jewelry, gloves, handbags, hats, belts, scarves, watches, sunglasses, pins, stockings, bow tie, leg warmer, leggings, necktie, suspenders, and tights.

Accessories add color, style and class to an outfit, and create a certain look, but they may also have practical functions. Handbags are for carrying, hats protect the face from weather elements, and gloves keep the hands warm[①].

Many accessories are produced by clothing design companies. However, there has been an increase in individuals creating their own brand name by designing and making their own label of accessories.

Accessories may be used as external visual symbols of religious or cultural affiliation: Crucifixes, Islamic headscarves, skullcaps and turbans are common examples[②]. Designer labels on accessories are perceived as an indicator of social status[③].

- Rings are worn by both men and women and can be of any quality. Rings can be made of metal, plastic, wood, bone, glass, gemstone and other materials. They may be set with a "stone" of some sort, which is often a precious or semi-precious gemstone such as ruby, sapphire or emerald, but can also be of almost any material[④].
- A brooch is a decorative jewelry item designed to be attached to garments. It is usually made of metal, often silver or gold but sometimes bronze or some other materials.
- A bracelet is an article of jewelry which is worn around the wrist. Bracelets can be manufactured from leather, cloth, hemp, plastic or metal, and sometimes contain rocks, wood, and/or shells.
- Earrings are jewelry attached to the ear through a piercing in the earlobe or some other external part of the ear. Earrings are worn by both sexes. Earring components may be made of any kind of materials, including metal, plastic, glass, precious stones, beads, and other materials.
- A belt is a flexible band, typically made of leather or heavy cloth, and worn around the waist. A belt supports trousers or other articles of clothing, and it serves for style and decoration.
- A purse or handbag is often fashionably designed, typically used by women, and is used to hold items such as wallet, keys, tissues, makeup, a hairbrush, cellular device or personal digital assistant, feminine hygiene products, or other items[⑤].

● A glove is a type of garment which covers the hand. Gloves have separate sheaths or openings for each finger and the thumb; if there is an opening but no covering sheath for each finger they are called "fingerless gloves". Fingerless gloves with one large opening rather than individual openings for each finger are sometimes called gauntlets. Gloves which cover the entire hand but do not have separate finger openings or sheaths are called mittens.

● A hat is a head covering. It may be worn for protection against the elements, for religious reasons, for safety, or as a fashion accessory.

● A scarf is a piece of fabric worn on or near the head or around the neck for warmth, cleanliness, fashion or for religious reasons.

● A watch is a timepiece that is made to be worn on a person. The term now usually refers to a wristwatch, which is worn on the wrist with a strap or bracelet.

● Sunglasses or sun glasses are forms of protective eyewear that usually enclose or protect the eye pupil in order to prevent strong light and UV rays from penetrating[6].

● A pin is a device used for fastening objects or material together. It is usually made of steel, or on occasion copper or brass. It is formed by drawing out a thin wire, sharpening the tip, and adding a head.

● A stocking, sometimes referred to as hose, is a close-fitting, variously elastic garment covering the foot and lower part of the leg. Stockings vary in color and transparency.

● Leg warmers are coverings for the lower legs, similar to socks but thicker and generally footless. Leg warmers were originally dancewear worn by ballet and other classic dancers.

● Leggings are any of several sorts of fitted clothing to cover the legs. Originally leggings were two separate garments, one for each leg.

● The necktie (or tie) is a long piece of cloth worn around the neck or shoulders, resting under the shirt collar and knotted at the throat. Variants include the bow tie, ascot tie, bolo tie, and the clip-on tie.

● Suspenders or braces are fabric or leather straps worn over the shoulders to hold up trousers.

● Tights are a kind of cloth leg garment, most often sheathing the body from about the waist to the feet with a more or less tight fit.

● A necklace is an article of jewellery which is worn around the neck. Necklaces are frequently formed from a metal jewellery chain, often attached to a locket or pendant.

图 28 Accessory design is an important part of fashion

Words and Phrases

glove [glʌv] *n*. 手套
handbag [ˈhændbæg] *n*. 女用手提包,小旅行袋,旅行包
scarf [skaːf] *n*. 围巾,披巾,披肩,头巾,领巾,腰巾
watch [wɔtʃ] *n*. 手表
sunglass [ˈsʌnˌglæs] *n*. 太阳镜
stocking [ˈstɔkiŋ] *n*. 长袜
bow tie 蝴蝶结
leg warmer 暖腿套,护腿
leggings [ˈlegiŋz] *n*. 儿童护腿套裤,开裆裤,裤绑腿,袜统
necktie [ˈnekˌtai] *n*. 领带,领结
ascot tie 阿司阔领带,领巾式领带
bolo tie 保罗领带,饰扣式领带
clip-on tie 夹式领带,夹式领结
necklace [ˈneklis] *n*. 项链
suspenders [səˈspendəz] *n*. 背带,吊带
tights [taits] *n*. 裤袜,紧身衣裤
outfit [ˈautfit] *n*. 服装,全套服装
affiliation [əˌfiliˈeiʃən] *n*. 联系,归属,关系
Crucifix [ˈkruːsəˌfiks] *n*. 耶稣受难像,十字架
Islamic headscarf 伊斯兰头巾
skullcap [ˈskʌlkæp] *n*. 无檐便帽,瓜皮帽
turban [ˈtɜːbən] *n*. 女式头巾,穆斯林头巾
semi-precious gemstone 半宝石的
ruby [ˈruːbi] *n*. 红宝石
sapphire [ˈsæfaiə] *n*. 蓝宝石,人造白宝石
emerald [ˈemərəld] *n*. 翡翠,绿宝石
brooch [bruːtʃ] *n*. 胸针,领针,扣花,胸饰
bracelet [ˈbreislit] *n*. 手镯,腕饰物
pierce [piəs] *vt*. 刺穿,刺破
earlobe [ˈiəˌləub] *n*. 耳垂
belt [belt] *n*. 腰带
dinner jacket 小礼服,晚宴茄克礼服,晚礼服
purse [pəːs] *n*. 小钱袋,女用小包
tissue [ˈtisjuː] *n*. 薄纱,餐巾纸
hairbrush [ˈheəˌbrʌʃ] *n*. 发刷
cellular device 手机
hygiene product 卫生用品
gauntlet [ˈgɔːntlit] *n*. 长手套,防护手套
mitten [ˈmitn] *n*. 连指手套
sheath [ʃiːθ] *n*. 鞘,套,外层覆盖(物),外包物
locket [ˈlɔkit] *n*. (装有照片或贵重金属纪念品的)项链
pendant [ˈpendənt] *n*. 垂饰,挂件;有垂饰的项链

Notes

① Accessories add color, style and class to an outfit, and create a certain look, but they may also have practical functions. Handbags are for carrying, hats protect the face from weather elements, and gloves keep the hands warm.
配饰给服装增添了色彩感、时尚感和品味,形成了某种特定的外观风格,但是,他们也具备实用功能。手袋可以用于携带物品,恶劣的气候下帽子可以保护面部,手套可以保暖。

② Accessories may be used as external visual symbols of religious or cultural affiliation: Crucifixes, Islamic headscarves, skullcaps and turbans are common examples.
配饰可能被视为宗教或文化起源的外部视觉符号:耶稣十字架、伊斯兰头巾、牧师的无檐便帽、穆斯林缠头巾都是常见的例子。

③ Designer labels on accessories are perceived as an indicator of social status.
配饰上的设计师标签被视为是社会地位的标识。

④ Rings are worn by both men and women and can be of any quality. Rings can be made of metal, plastic, wood, bone, glass, gemstone and other materials. They may be set with a "stone" of some sort, which is often a precious or semi-precious gemstone such as ruby, sapphire or emerald, but can also be of almost any material.
戒指品质多样,男女均可佩戴。金属、塑料、木材、骨头、玻璃、宝石或其他材料均可以制作成戒指。戒指可能会被装置某种"石头",通常是贵重的宝石或半宝石,例如红宝石、蓝宝石或翡翠,也可以使用其他任何材料。

⑤ A purse or handbag is often fashionably designed, typically used by women, and is used to hold items such as wallet, keys, tissues, makeup, a hairbrush, cellular device or personal digital assistant, feminine hygiene products, or other items.
手袋或手提包通常设计得很时尚,多为女性携带,用于放置钱包、钥匙、纸巾、化妆品、发梳、手机或私人数码产品、女性卫生用品等。

⑥ Sunglasses or sun glasses are forms of protective eyewear that usually enclose or protect the eye pupil in order to prevent strong light and UV rays from penetrating.
太阳镜是保护眼睛的装置,用于保护瞳孔防止被强光和紫外线损伤。

Discussion Questions

1. List the accessory which you wear today and talk about their color and shape according to your style with your classmate.

EXTENSIVE READING

AREAS OF FASHION DESIGN

Many professional fashion designers start off by specializing in a particular area of fashion. The smaller and the more specific the market, the more likely a company is to get the right look and feel to their clothes. It is also easier to establish oneself in the fashion industry if a company is known for one type of product, rather than several products. Once a fashion company becomes established (that is, has regular buyers and is well-known by both the trade and the public), it may decide to expand into a new area. If the firm has made a name for the clothes it already produces, this helps to sell the new line. It is usually safest for a company to expand into an area similar to the one it already

图 29 kids apparels

knows. For example, a designer of women's sportswear might expand into men's sportswear. The chart below shows the areas in which many designers choose to specialize.

Area	Brief	Market
Women's Day wear	Practical, comfortable, fashionable	Haute couture, ready-to-wear, mass market
Women's Evening wear	Glamorous, sophisticated, apt for the occasion	Haute couture, ready-to-wear, mass market
Women's Lingerie	Glamorous, comfortable, washable	Haute couture, ready-to-wear, mass market
Men's Day wear	Casual, practical, comfortable	Tailoring, ready-to-wear, mass market
Men's Evening wear	Smart, elegant, formal, apt for the occasion	Tailoring, ready-to-wear, mass market
Girls' Wear	Pretty, colorful, practical, washable, inexpensive	Ready-to-wear, mass market
Teenage Wear	Highly fashion-conscious, comfortable, inexpensive	Ready-to-wear, mass market
Sportswear	Comfortable, practical, well-ventilated, washable	Ready-to-wear, mass market
Knitwear	Right weight and color for the season	Ready-to-wear, mass market
Outerwear	Stylish, warm, right weight and color for the season	Ready-to-wear, mass market
Bridal wear	Sumptuous, glamorous, classic	Haute couture, ready-to-wear, mass market
Accessories	Striking, fashionable	Haute couture, ready-to-wear, mass market

Words and Phrases

teenage wear　少年装
bridal wear　婚礼服
glamorous [ˈglæmərəs] *adj*. 迷人的
apt [æpt] *adj*. 适当的，切题的
smart [smaːt] *adj*. 漂亮的
pretty [ˈpriti] *adj*. 优雅的，潇洒的
formal [ˈfɔːməl] *adj*. 正式的
ready-to-wear　现成的服装，高级女装成衣
elegant [ˈeligənt] *adj*. 优雅的，优美的
fashion-conscious　时尚感的
practical [ˈpræktikəl] *adj*. 实用的
well-ventilated　透气的
stylish [ˈstailiʃ] *adj*. 时尚的，时髦的
sumptuous [ˈsʌmptʃuːəs] *adj*. 豪华的，奢侈的
striking [ˈstraikiŋ] *adj*. 引人注目的，吸引人的

Chapter 4 GARMENT CONSTRUCTIONS / 服装结构

LESSON 15 PRINCIPLES OF PATTERN MAKING / 服装纸样设计原理

Basic principles are common to many pattern pieces, these should be considered before one begins.

Seam Lines

A pattern piece can be cut across vertically, horizontally, diagonally, with curved lines etc. When the sections are joined the pattern piece will have a seam, but the basic shape remains the same. Dart shaping can be moved to seam lines so that the shaping remains but the dart disappears①.

Shape

A garment can fit closely to the figure, be semi-fitting or easy fitting in shape. This is achieved by using the blocks with or without shaping. Some examples of changes of pattern shape are widening the outline... inserting extra body ease; hidden shapes... adding pleats and godets; puff and bell shapes... adding width to the design by tucks or gathers; cone shapes... widening the hem line only②.

Adding Pieces

When adding pockets, peplums, panels, flaps, etc. consider carefully the balance of the design.

Body Movement

In more advanced pattern cutting parts of bodices are added to sleeves. When working these designs always be aware that the body must be able to move. It is only on wide, full garment that very simple shapes can be used.

Beautiful Shapes

It is always necessary to have good lines and shapes. When cutting intricate patterns small amounts of the basic block may be lost or small parts added. How much one can do this depends on one's skill and experience. That is why it is so difficult to cut the subtle shapes achieved by our top designers③.

Cutting individual garments gives designers much more freedom; they are not restricted by the price limits and the fabrics used in mass production.

Manufacturers require finished patterns to have seam allowances added. Some require their designers to adapt patterns from blocks that already include the seam allowance[④].

Seam allowance widths vary with the type of manufacture and garment.

- Basic Seams e.g. side seams, style seams...1 to 1.5 cm.
- Enclosed Seams e.g. collars, cuffs...0.5 cm.
- Hems depth depends on shape and finish...1 to 5 cm.
- Decorative seams usually require more seam allowance.

Fabrics that fray easily will require wider turnings especially around facings and collars. The width of the seam allowance must be marked on the pattern by lines or notches[⑤].

No seam allowance is required on a fold line.

It is important that seam allowances added to the pattern are accurate and clearly marked.

Toile Patterns

It is necessary to add seam allowances at this stage, they can be marked directly on the calico.

Industrial Patterns

The seam lines are not marked on these patterns. The seam allowance is usually stated in the making up specification and only varying seam allowances will be marked by notches[⑥].

To enable the garment to be made up correctly the following instructions must be marked on the pattern:

- The name of each piece.
- Centre back and centre front.
- The number of pieces to be cut.
- Folds.
- Balance marks...these are used to make sure pattern pieces are sewn together at the correct points[⑦].
- Seam allowances... these can be marked by lines round the pattern or notches at each end of the seam. If a pattern is nett (has no seam allowance), mark this clearly on the pattern.
- Construction lines... these include darts, buttonholes, pocket placings, tucks, pleat lines, decorative stitch lines. These lines are marked on the pattern or shown by punch holes.

图 30 Traditional tag-board pattern storage in a technical design room

- Grain lines... to achieve the effect you require, you must understand the principle of placing a pattern on the correct grain of the fabric. Mark the grain line with an arrow. Mark the grain lines on the working pattern before it is cut up into sections. Once in pieces it can be difficult to find the correct grain on complicated pattern sections.
- Pattern size.
- Style No.

Words and Phrases

seam line 接缝线;沿缝
construction line 结构线
shape [ʃeip] *n*. 形状,外形;(人体的)特有形状;体形,身段
dart [dɑːt] *n*. (衣服上的)缝褶,省道,缉省
semi-fitting 半紧身的
easy fitting 穿着舒适的
block [blɔk] *n*. 原型,裁剪样板
ease [iːz] *n*. 松份;放松,放宽;拔开;舒适
pleat [pliːt] *n*. 打褶,褶裥
godet [gəu'det] *n*. 三角布(用以放宽衣裙下摆),裆布
puff [pʌf] *n*. 尖角布;整片;皱褶;胖褶
bell [bel] *n*. 喇叭口;降落伞衣;钟形物
tuck [tʌk] *n*. 横裥,活褶,塔克 *v*. 打褶,打裥
gather ['gæðə] *n*. 褶裥,碎褶 *v*. 打褶裥
cone [kəun] *n*. 锥体,锥形
peplum ['pepləm] *n*. 腰褶
panel ['pænəl] *n*. 嵌条,饰条,布块
flap [flæp] *n*. 前门襟;袋盖;帽边
restrict [ris'trikt] *v*. 限制,限定,约束
seam allowances 缝接允差,缝合允许量,缝头,缝份
basic seam 基本线,基本缝纫结构线
enclosed seam 封边缝,止口缝
hem [hem] *n*. 贴边,卷边,下摆,脚口折边,缝边下缘
fray [frei] *v*. 磨损
facing ['feisiŋ] *n*. 挂面,贴条,贴边
notch [nɔtʃ] *n*. 刀口,刀眼 *v*. 打刀眼
foldline 折边线,对折线,折叠线
toile pattern 样衣纸样
industrial pattern 工业用纸样
calico ['kælikəu] *n*. [英]白棉布;[美]印花布
notch [nɔtʃ] *n*. 切口
nett pattern 净样
pocket placing 袋位
pleat line 裥位线
decorative stitch line 装饰缉线
punch hole 孔位记号;穿孔
grain line 经向线
style No. 款式号码

Notes

① A pattern piece can be cut across vertically, horizontally, diagonally, with curved lines etc. When the sections are joined the pattern piece will have a seam, but the basic shape remains the same. Dart shaping can be moved to seam lines so that the shaping remains but the dart disappears.
衣片可以沿纵向、横向、斜向直裁开,也可以用弧线分割。当裁开的衣片缝合时,基本形状不会发生变化,但是服装会出现拼缝线。省道也可被转移至拼缝线中,结果

是服装造型未变，但是省道消失了。

② A garment can fit closely to the figure, be semi-fitting or easy fitting in shape. This is achieved by using the blocks with or without shaping. Some examples of changes of pattern shape are widening the outline... inserting extra body ease; hidden shapes... adding pleats and godets; puff and bell shapes... adding width to the design by tucks or gathers; cone shapes...widening the hem line only.

服装可以是紧身的、半紧身或宽松的，这取决于如何处理原型。例如，可以通过加入额外的松量来放大尺寸，设计褶裥来收身，增加塔克和碎褶来形成喇叭造型，或仅仅增加下摆的围度形成锥形造型。

③ It is always necessary to have good lines and shapes. When cutting intricate patterns small amounts of the basic block may be lost or small parts added. How much one can do this depends on one's skill and experience. That is why it is so difficult to cut the subtle shapes achieved by our top designers.

优美的线条和形状是必需的。当裁剪复杂的纸样时，基本原型会舍弃某些细部结构，或另外增加小块衣片，如何取舍取决于个人的技术和经验，这就是裁剪顶级设计师的纸样非常困难的原因。

④ Manufacturers require finished patterns to have seam allowances added. Some require their designers to adapt patterns from blocks that already include the seam allowance.

生产商要求大生产的纸样必须有缝份。有些生产商会要求设计师修改那些根据毛样原型绘制的纸样。

⑤ Fabrics that fray easily will require wider turnings especially around facings and collars. The width of the seam allowance must be marked on the pattern by lines or notches.

易磨损的面料要求在挂面和领子上有更宽的翻折量，缝份的宽度必须用直线和刀眼在纸样上标示清楚。

⑥ The seam lines are not marked on these patterns. The seam allowance is usually stated in the making up specification and only varying seam allowances will be marked by notches.

净缝线不标在(工业)纸样上。缝份通常会在生产工艺单中标明，并且只有可调节的缝份会被刀眼标记。

⑦ Balance marks...these are used to make sure pattern pieces are sewn together at the correct points.

对位记号：帮助确定纸样在正确位置拼合。

Discussion questions

1. What are the principles of pattern making?

EXTENSIVE READING

GARMENT CONSTRUCTION TERMS

Bodice

The portion of the garment above the waist.

Gores

Gores are vertical pieces of fabric that shape the fabric to the body when sewn together.

Seams

Seam is the name given to the sides of pattern pieces which are stitched together to form the garment.

Facing

An extra piece of fabric that covers raw edges at the neckline, armhole, and other garment openings.

Fasteners

Items used to hold a garment closed, such as buttons, snaps, hooks and eyes, and zippers.

Darts

A dart is a wedge-shaped piece of fabric that is used to stitch out excess fabric where the body is most slender and tapers to nothing where the body is fullest.

Gathers

Soft, unstitched folds of fabric used to control fullness.

Hem

Fabric folded back at the bottom edge of a garment to finish off the edge.

Interfacing

An extra layer of fabric placed between the garment and the facing to add shape or body.

Style Ease

Style ease is extra fabric which is distributed over the curves of the body by gathering or pleating fabric into a seam.

Lapel

Part of the garment that turns back below the collar.

Lining

An extra layer of fabric used to prevent stretching and to finish off the inside of the garment.

Placket

A garment opening finished with a strip of fabric or zipper.

Yokes

Yokes are horizontal divisions in a garment used for styling and fitting.

Pleat

A wider fold of fabric pressed or stitched flat.

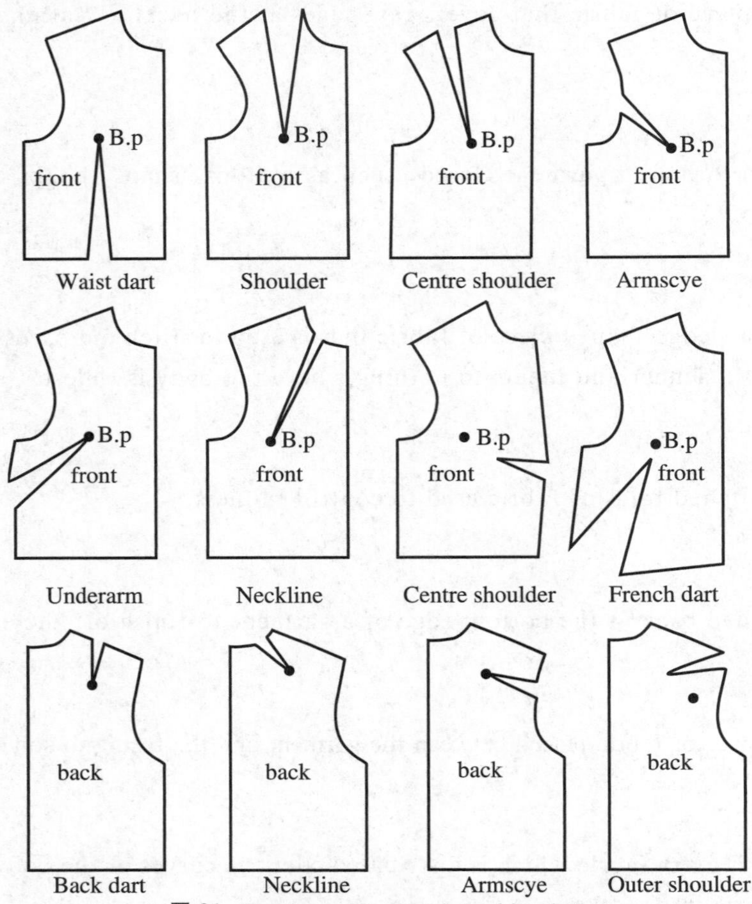

图 31 Positions of bust and shoulder darts

Trim

Some form of decoration added to a garment, such as braid, lace, or rickrack.

Tuck

Narrow fold of fabric stitched flat.

Waistline

A horizontal seam attaching the upper and lower sections of a garment; it can be located above, at, or below the natural waistline.

Words and phrases

gore [gɔː] *n*. 衣片，拼块，三角布
fullness [ˈfulnis] *n*. 丰满度
raw edge 毛边
neckline [ˈneklain] *n*. 领围线
armhole [ˈaːmhəul] *n*. (AH)袖窿
wedge-shaped 楔形
slender [ˈslendə] *adj*. 细瘦的，苗条的
fish-eye dart 鱼眼形省道
commercial placement 商业纸样的处理方法
placket [ˈplækit] *n*. 开口，衣袋
yoke [jəuk] *n*. 过肩，覆肩，育克，约克
hip line 臀围线
rickrack [rikˈræk] *n*. 波曲形花边，荷叶边
waist dart 腰省
armscye [aːmˈziː] *n*. 袖窿
underarm [ˈʌndəraːm] *adj*. 手臂下的，腋下的；*n*. 腋点
french dart 曲线省，刀背缝

LESSON 16 FLAT PATTERN-MAKING / 平面纸样裁剪

Flat pattern cutting is the customary method used in the fashion industry. Patterns are developed in the retailer or brand's standard size, which is usually a UK 12 (European 38/40 or US 8/10) for womenwear.

Basic blocks are adapted in a first draft to reflect the required fit and relevant details for a garment design by tracing onto transparent paper or by using a tracing wheel on card. Students are sometimes taught to cut patterns using fifth-scale blocks for speed and economy, but in practice in industry full-scale blocks are always used[①]. The edges of the blocks on the first draft are where the seams of the garment will meet, so a working pattern is traced off with seam allowances varying from 0.5cm to 1.5cm added around the edges to enable the product to be sewn together.

The size of the seam allowances depends on the type of machinery and fabric to be used, indicated by adding small notches near the corners of the pattern. Pattern cutters need to be very neat and thorough with attention to detail. They work with precise measurements and if they do not use the correct seam allowance, the fit and quality of the production garments can be adversely affected, resulting in high returns and poor

sales[②]. Pattern cutters need a thorough understanding of the construction of garments.

Various aspects of the block can be adapted to reflect the required style, including manipulating darts to affect the fit and styling, adding seams as design details, adding collars or revers to tops, adding fullness to skirts with gores, godets or pleats and reducing or lengthening hems[③]. When the styling and design details on the pattern are finished, pattern pieces for facings and interfacings, which help with the finish and construction of the garment, can be added if necessary.

Patterns require the following annotation:

- Reference number or name of a garment style.
- Garment size.
- Centre-front (CF) or centre-back line (CB).
- Grainline.
- Folds.
- Balance marks (notches to ensure back and front pieces are sewn together correctly).
- Number of pieces (to ensure none are missing).
- The name of the piece, as some parts can look very similar.
- Construction lines, e.g. darts, buttonholes and pocket positions.

The "grain" of the fabric lies parallel to the selvedge (the finished edges of the cloth where it was attached to the loom). On the pattern the grainline indicates the direction of the grain, to explain how the pattern piece should be laid on the fabric. Designers may intentionally require parts of a garment, such as a waistband, to be placed "on the bias", i.e. diagonally on the fabric to take advantage of the stretch properties this offers[④]. It is important to organize the patterns well as one of the marks of a good pattern cutter is that a colleague can easily find patterns they require if the pattern cutter is out of the office.

Knitwear pattern cutting tends to be based on relatively simple shapes as there is a lot of flexibility in the fabric's stretch properties. "Cut-and-sew" knitwear is made from pieces of knitted fabric called "blanks" which are cut to shape then sewn together with over-lockers. The garments are then finished at the neck, cuffs and hem, often using knitted ribs. Fully-fashioning is a traditional and high quality method of making knitwear with all of the pieces knitted to the finished shape, so no fabric cutting is required and there is minimal wastage.

图 32 A patternmaker is drawing the seam allowance on the pattern

Words and phrases

draft [drɑːft] *n*. 样板,原图,草案

fifth-scale block　1∶5 纸样

full-scale blocks　原尺寸,实际尺寸

neat [niːt] *adj*. 整洁的,匀称的

measurement [ˈmeʒəmənt] *n*. 围度，测量，尺寸

rever [ˈrivə] *n*. 翻边，翻领，翻袖

tops [tɔps] *n*. 上装

annotation [ˌænəuˈteiʃən] *n*. 注解，注释

selvedge [ˈselvidʒ] *n*. 织边，布边，边缘

waistband [ˈweistbænd] *n*. 腰带

cut-and-sew　裁剪成型（相对于全成型针织服装成衣法）

over-locker　拷边机，包缝机

knitted rib　针织罗纹

fully-fashioning　全成型的

wastage [ˈweistidʒ] *n*. 损耗

Notes

① Basic blocks are adapted in a first draft to reflect the required fit and relevant details for a garment design by tracing onto transparent paper or by using a tracing wheel on card. Students are sometimes taught to cut patterns using fifth-scale blocks for speed and economy, but in practice in industry full-scale blocks are always used.
基础原型首先经过调整来获得款式所需的放松量和相关的结构细节，并通过透明纸或描线轮转移到卡纸上。为了提高打样速度和从节约角度考虑，学生常常被建议使用1∶5的小样，实际上，工厂一直使用实样。

② The size of the seam allowances depends on the type of machinery and fabric to be used, indicated by adding small notches near the corners of the pattern. Pattern cutters need to be very neat and thorough with attention to detail. They work with precise measurements and if they do not use the correct seam allowance, the fit and quality of the production garments can be adversely affected, resulting in high returns and poor sales.
缝份的尺寸取决于缝纫设备和缝纫面料，由靠近纸样边角的刀眼表示。纸样裁剪师需要麻利而精确地处理细节。他们精确测量尺寸，如果没有采用正确的缝份，服装的合体性和生产质量将会受到不良影响，并导致高返工率和滞销。

③ Various aspects of the block can be adapted to reflect the required style, including manipulating darts to affect the fit and styling, adding seams as design details, adding collars or revers to tops, adding fullness to skirts with gores, godets or pleats and reducing or lengthening hems.
原型的许多地方能被修改以此获得所需的造型，包括修改省道来调节合体度和造型，增加拼缝设计细节，为上装增加立领或翻领，利用褶、裥或皱褶来增加裙子的丰满度，修改下摆的长短等。

④ Designers may intentionally require parts of a garment, such as a waistband, to be placed "on the bias", i.e. diagonally on the fabric to take advantage of the stretch properties this offers.
设计师可能会故意安排服装的一部分斜裁，例如腰带部分，利用面料斜丝绺提供的拉伸性。

Discussion questions

1. What products might be better served by draping pattern making methods rather than flat-pattern methods?
2. Compare the garment made from draping to the garment made from flat-pattern.

EXTENSIVE READING

MEASUREMENT FOR BASIC BLOCKS

Block patterns are foundation patterns constructed to fit the body measurements of an average figure of one of the size groups (10, 12, 14, etc.). The blocks include the basic amount of ease for the function of the garment. Underwear requires less ease than over-garments.

The designer completes a pattern in a sample size, usually size 10 or 12. The design is made up into a sample garment. When the design is accepted, the garment is graded into the remaining sizes required by the buyer.

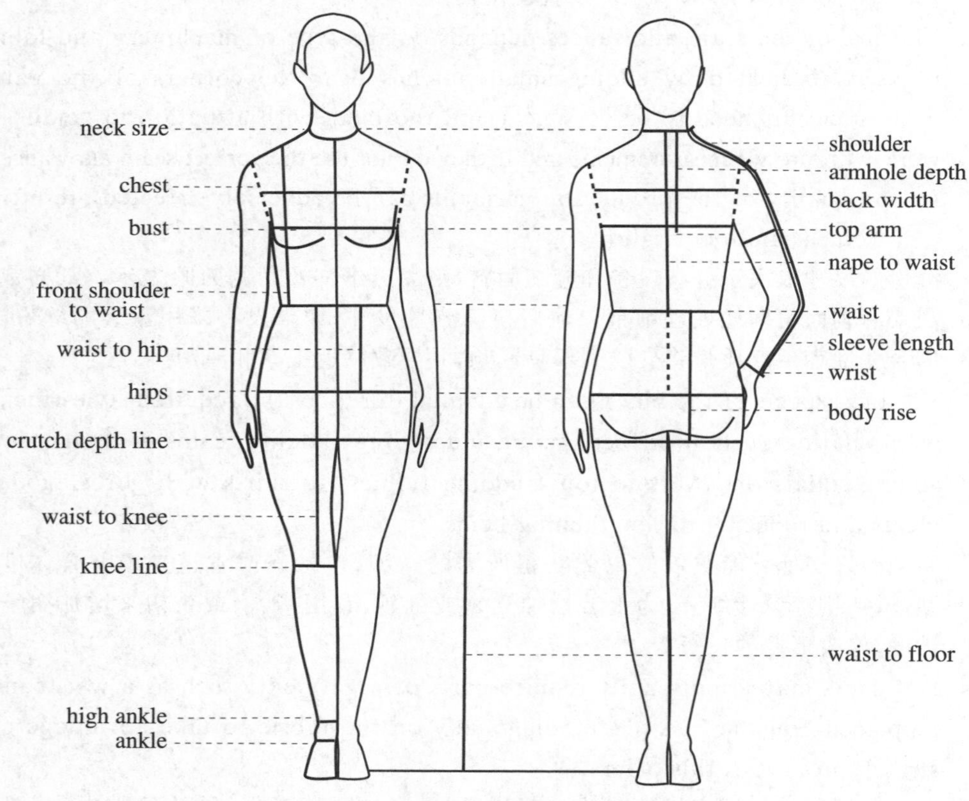

图 33 Standard body measurements

Women of Medium Height 160cm～170cm (5ft 2½ in—5ft 6½ in)							
Size	10	12	14	16	18	20	22
Bust	82	87	92	97	102	107	112
Waist	62	67	72	77	82	87	92
Hips	87	92	97	102	107	112	117
Back Width	33	34.2	35.4	36.6	37.8	39	40.2
Chest	30.5	32	33.5	35	36.5	38	39.5
Shoulder	11.9	12.2	12.5	12.8	13.1	13.4	13.7
Neck Size	35.6	36.8	38	39.2	40.4	41.6	42.8
Dart	6.4	7	7.6	8.2	8.8	9.4	10
Top Arm	26.4	28	29.6	31.2	32.8	34.4	36
Wrist	15.5	16	16.5	17	17.5	18	18.5
Ankle	23.4	24	24.6	25.2	25.8	26.4	27
High Ankle	20.4	21	21.6	22.2	22.8	23.4	24
Nape to Waist	39.5	40	40.5	41	41.5	42	42.5
Front Shoulder to Waist	39.5	40	40.5	41.3	42.1	42.9	43.7
Armhole Depth	20.5	21	21.5	22	22.5	23	23.5
Waist to Knee	58	58.5	59	59.5	60	60.5	61
Waist to Hip	20.3	20.6	20.9	21.2	21.5	21.8	22.1
Waist to Floor	103	104	105	106	107	108	109
Body Rise	27.3	28	28.7	29.4	30.1	30.8	31.5
Sleeve Length	57.3	58	58.7	59.4	60.1	60.8	61.5
Sleeve Length (Jersey)	51.3	52	52.7	53.4	54.1	54.8	55.2

Words and Phrases

basic block　原型

bust [bʌst] *v*. 胸围，胸部

chest [tʃest] *n*. 前胸

neck [nek] *n*. 颈部，领部

wrist [rist] *n*. 腕，腕部

ankle [ˈæŋkl] *n*. 踝骨，脚脖

nape [neip] *n*. 颈背，后颈

rise [raiz] *n*. 立裆，直裆，裆

high ankle　上踝围

nape to waist　后腰长

front shoulder to waist　前肩线至腰线长

armhole depth　袖窿深

waist to knee　腰线至膝围线长

waist to hip　腰线至臀围线长

waist to floor　腰线至地面长
body rise　裆深
sleeve length　袖长
sleeve length(jersey)　紧身衫袖长

LESSON 17　DRAPING / 立体裁剪

Fabric is often the source of inspiration for the creation of designs, but it is the draper's skillful hands that manipulate shapeless pieces of cloth into beautiful garments.

The first version of most garments is draped in muslin, because it is an economical fabric. Ideally, a garment should be draped in the design fabric or a substitute that is closely related in texture and weight. Although this may be too expensive for the manufacturer, it may be appropriate when draping for private clientele①. At completion, the drape should be critiqued for style-line placement, proportion, balance, and fit before removing it from the form or model.

After the critique, the pins are removed from the draped design. The marked style-lines and seams are trued. Truing a draped garment requires that all markings placed on the muslin indicating style-lines and seams be straightened and curved lines blended②. Measurements should be compared with those on the measurement chart. These steps are necessary if an accurate outline of the pattern shapes is to be achieved. The cloth patterns can be stitched and placed on the form, or model, for a test fit, or transferred to paper, cut, and stitched in the design fabric for fitting.

The beginning draper is often surprised to learn that the pinned drape that looked near perfect on the form can still have fitting problems when cut and stitched in the design fabric. This is to be expected for two reasons:

1. The design fabric, which has a different texture and weight from that of draped muslin, can result in a garment that hangs differently on the figure, thereby creating fitting problems.
2. Inaccurate marking and truing of muslin patterns that were then transferred to paper also cause fitting problems.

图 34　A designer is draping a dress with the muslin cloth on the form

Beginning drapers are greatly tempted to put fabric on the form to create wonderful designs, but draping is not that simple. Every accomplished artist knows that it takes hard work and determination to achieve perfection. Draping depends on controlling the straight and crosswise grainlines when manipulating the fabric on the form to create the desired design effect, and the balance of the garment. It also involves an understanding of the principles that guide the draper in choosing the correct draping techniques required by the design③.

Finally, the draper must have knowledge of the characteristics of fabrics and the ability to choose a fabric that will be compatible with the design. The identification of fitting problems and their solution is an ongoing learning experience. To go forward, remember that perseverance is the key, as is the love for draping beautiful garments.

Garments that are designed for the retail market require production patterns. The garment may go through several test samples to assure that the fit and patterns are perfect before being released for production.

Designers who create one-of-a-kind garments for private clientele often drape in the design fabric. In some cases after the drape has been completed, the garment is removed and basted for a fitting. Adjustments are made, and the garment is stitched. It is ready for wear without the creation of a pattern for the design because it will not be made again.

Words and Phrases

muslin [ˈmʌzlin] *n*. 平纹细布,麦斯林纱
economical [ˌikəˈnɔmikəl] *adj*. 节俭的,经济的,合算的
clientele [ˌklaiənˈtel] *n*. 客户
critique [kriˈtiːk] *n*. 评论,评价
style-line 造型线,款式结构线
true [truː] *vt*. 调整,修形
crosswise [ˈkrɔsˌwaiz] *adv*. 横向的
ongoing [ˈɔnˌgəuiŋ] *adj*. 进行的
perseverance [ˌpəːsiˈviərəns] *n*. 坚持
test fit 裁剪前的试样
one-of-a-kind 独一无二的
baste [beist] *v*. 用长针脚粗缝

Notes

① Ideally, a garment should be draped in the design fabric or a substitute that is closely related in texture and weight. Although this may be too expensive for the manufacturer, it may be appropriate when draping for private clientele.
理想状态是使用设计指定的面料或肌理和重量都相似的替代品进行立体裁剪。尽管就制作者而言材料太贵,但是对于私人订单是适合的。

② Truing a draped garment requires that all markings placed on the muslin indicating style-lines and seams be straightened and curved lines blended.
描实立裁服装要求将坯布上所有的标识描成(流畅的)直线形和弧线形的结构线。

③ Draping depends on controlling the straight and crosswise grain-lines when manipulating the fabric on the form to create the desired design effect, and the balance of the garment. It also involves an understanding of the principles that guide the draper in choosing the correct draping techniques required by the design.
立体裁剪依靠对放置于人台上的面料的经纬向的控制,来获得理想的设计效果和协调的搭配。引导制作者正确立裁的是对(服装)结构原理的理解和掌握。

Discussion questions

1. What is fit and how is it achieved?

EXTENSIVE READING

PROTOTYPING

The prototype is the first critical opportunity for the product development team to see an actual garment style and try it on a fit model. This opportunity to evaluate the aesthetic of the style on the human body is an important step, especially for fashion garments. It is also an opportunity to see if the style meets the company's standards for quality. For basic sportswear, it is also important in refining the fit of the garment. Merchandisers must strive to maintain consistency in the fit of their company's products. Apparel brand loyalty is often based upon fit. This is especially critical in underwear, foundations, jeans, dress shirts, and tailored clothing.

There are two primary methods for creating prototypes-draping and flat pattern. Draping is usually used by couture and ready-to-wear designers who create garments for the designer price range. This process of cutting, shaping, and draping fabric on a dress form or mannequin allows the designer to evaluate the way a fabric performs when it is shaped to conform to a three-dimensional shape.

Flat pattern is a process that involves altering basic pattern blocks or slopers to achieve the desired silhouette. It is more frequently used in ready-to-wear apparel companies to create prototypes. Slopers or blocks are sets of patterns for each basic garment type produced by the company. They have been refined over a period of many seasons to provide the exact fit desired by the target market consumer. Today's CAD pattern design systems (PDS) are the most efficient method of flat pattern design. They achieve a high degree of accuracy and speed when creating new patterns for a prototype. They also easily interface with single-ply, automated sample cutters to allow the designer to quickly construct a prototype to check proper fit. After the fit sample is tested, pattern adjustments can be made quickly and easily.

With the proliferation of global sourcing, the sample-making process is often transferred to a manufacturer who may be located halfway around the world. The cost and time required to make and deliver prototypes from global sourcing partners make controlling the product development process a challenge. It requires that merchandisers monitor the initial phases of product development carefully in order to weed out any potential styles that may not fulfill the needs of the line plan. Accurate communications of prototype requirements is another critical issue for merchandisers. Computer-based PIM systems provide a consistent, accurate method of creating and communicating prototype requests, whether they are transmitted to a factory in the United States or to a sourcing contractor in South America or Southeast Asia.

The following data are critical for a prototype request:

- Company name.
- Season and date.

- Style code.
- Product identification.
- Prototype identification code.
- Detailed description code.
- Technical drawing of garment.
- Sample pattern identification codes.
- Sample size measurements with tolerances.
- Fabric description.
- Bill of materials.
- Cutting instructions.
- Sewing construction details.
- Labeling instructions.
- Finishing instructions.
- Packaging instructions.
- Target price.

A prototype also can be made using a form that is an exact replica of the company's fit model. When a contractor makes the prototypes the contractor will be supplied with a copy of the form. After the prototype has been sewn, it is placed on the contractor's form and a digital video is made. Using software that converts the 2-D video to 3-D, one can rotate the prototype 360 degrees, as well as zoom in and out to inspect quality. When these images are transmitted electronically from the producer of the prototype to the product development staff, adjustments can be made immediately. This increases efficiency and saves both time and money.

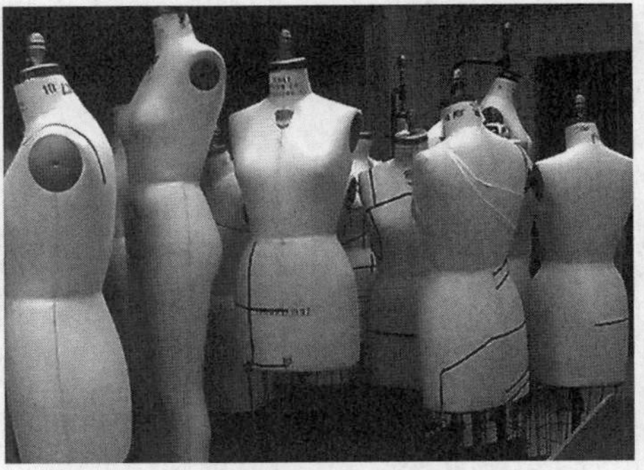

图 35 Dress forms come in a variety of shapes for different styling needs

Words and Phrases

prototype [ˈprəutətaip] *n.* 原型，模型，样板；标样，样品

strive to 努力，致力于……

based upon 基于……

mannequin [ˈmænikin] *n.* 人体模特，橱窗模特

sloper [ˈsləupə] *n.* 服装尺寸样板

single-ply 单层

proliferation [prəuˌlifəˈreiʃən] *n.* 增生，扩散

initial phases 初始阶段

weed out 清除

bill of materials 材料单

instruction [inˈstrʌkʃən] *n.* 说明书，细则

replica [ˈreplikə] *n.* 副本

contractor [kənˈtræktə] *n.* 承建商

zoom in 放大

zoom out 缩小

LESSON 18 TERMINOLOGY AND PATTERN LANDMARKS / 服装结构术语与标识

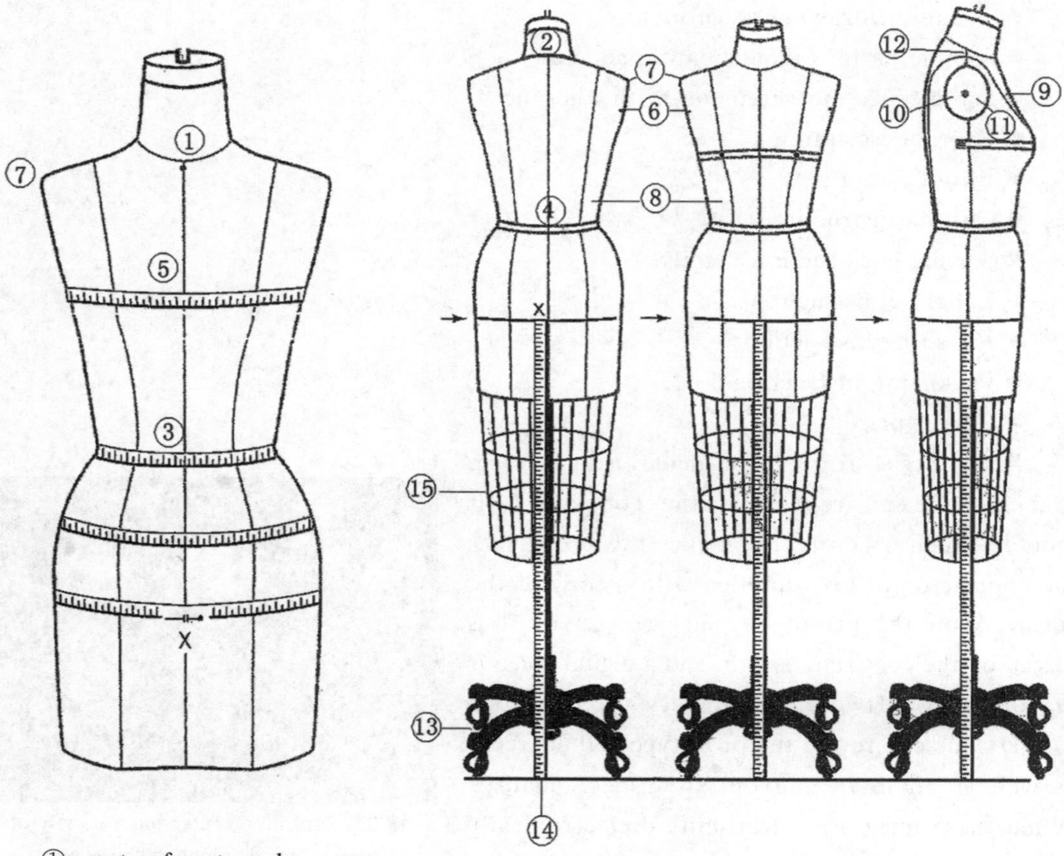

① center front neck
② center back neck
③ center front waist
④ center back waist
⑤ center front bust level (between bust points)
⑥ mid-armhole front/back
⑦ shoulder tip
⑧ side front/back (princess)
⑨ armhole ridge or roll line
⑩ plate screw
⑪ armhole plate
⑫ shoulder at neck
⑬ roller wheels
⑭ height pedal

⑮ cage

Symbol Key

CF=center front
CB=center back
BP=Bust point
SS=Side seam
SW=Side waist
SH=Shoulder
HBL=Horizontal balance line
SH-TIP=Shoulder tip

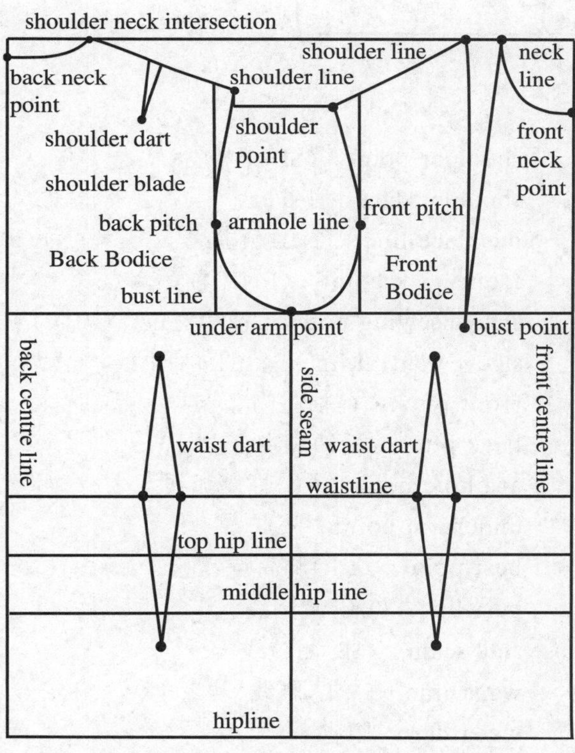

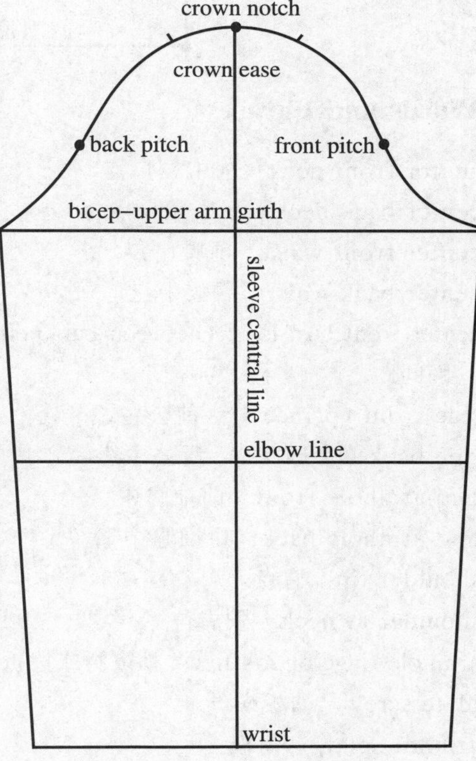

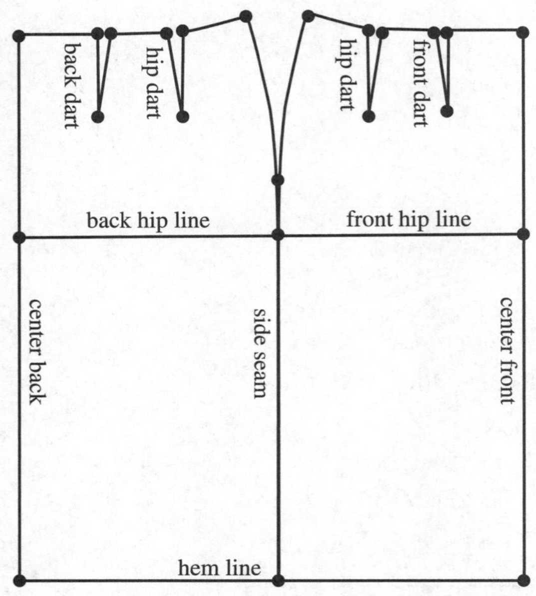

Words and Phrases

center front neck　前中领口点
center back neck　后中领口点
center front waist　前腰中点
center back waist　后腰中点
center front bust level (between bust points)　胸间距
side front (princess)　前公主线
side back (princess)　后公主线
mid-armhole front　前袖窿中
mid-armhole back　后袖窿中
shoulder tip　肩点
shoulder at neck　颈肩位
armhole ridge or roll line　袖窿脊或拨折线
plate screw　袖板螺丝
armhole plate　袖板
roller wheels　滚轮
height pedal　高度踏板
front bodice　前衣身
back bodice　后衣身
front central line　(FCL)前中心线
back central line　(BCL)后中心线
shoulder neck point　(SNP)侧颈点
shoulder point　(SP)肩点
shoulder dart　肩省
necklace line　(NL)颈围线
front neck point　(FNP)前颈窝点
sleeve sewing notch　袖片对位点(刀眼)
sleeve central line　(SCL)袖中线
front pitch　前袖窿吻合点
back pitch　后袖窿吻合点
armhole line　(AL)袖窿弧线
underarm point　腋点
bust point　(BP)胸高点,乳点
bust line　(BL)胸围线
side seam　(SS)侧缝线
waist line　(WL)腰线
waist dart　腰省
hip line　(HL)臀围线
top hip line　(THL)上臀线
middle hip line　(MHL)中臀围线
balance mark　对位记号
crown [kraun] *n.* 袖山
crown ease　袖山吃势
crown notch　袖山对位剪口

bicep-upper arm girth 上臂围 head size （HS）头围
elbow line （EL）肘围线 knee line （KL）膝围线

EXTENSIVE READING

GRADING

Initial samples are developed in a size which is in the middle of the full size range, so that the patterns can be graded up or down to all of the sizes required. When grading, a "nest" of patterns is created, with lines radiating from the smaller patterns to the surrounding larger sizes. The differences between the sizes are called increments and they vary depending on the part of the garment, the type of garment and the "grade rules" of the retailer or brand. There is usually an increment of 5cm in total circumference between each size in the chest, waist and hip measurements.

CAD/CAM systems tend to be used more frequently for the grading of production styles than for the development of sample patterns. A file of graded patterns can be utilized within compatible software to develop a lay plan for the most economical use of the fabric with the least possible wastage. In preparation for garment production, the bulk fabric is spread in multiple layers on a cutting table and the lay plan indicates the way in which it will be cut. A full-size printout of the lay plan is laid on top of the fabric which is cut to this template with an electric knife, resulting in bundles of individual garment pieces in varying sizes.

Circumference Grading

Grading rules are developed to accommodate size variations in girth, length, and width. Circumference grading indicates how many the garments are to increase in girth from on size to the next. It is important that the growth be distributed somewhat evenly around the body rather than being added in one place. These measurements are often the most changed from one size to another size, as individuals tend to reflect more differences in girth than in height.

Length Grading

Length grading refers to the measurement to be added to the length of garment pieces as sizes progress from one to the next. Most of these measurements are related to height. Length must be added in proportion to the body's natural growth. The typical amount of 1/4 inch in length per grade in the bodice should be distributed

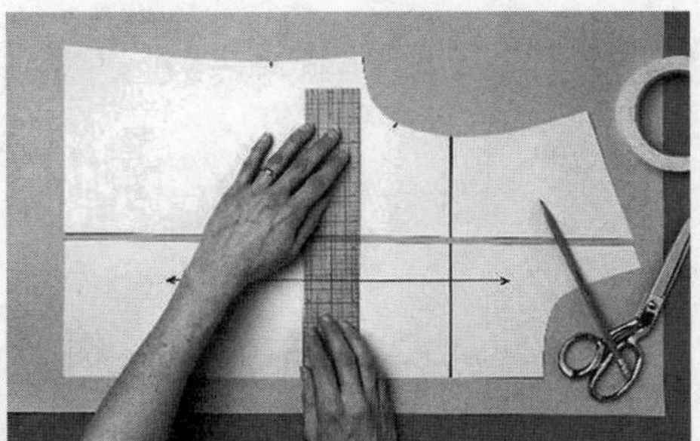

图 36 Pattern grading can be done either with a grading machine or by using a computer program

evenly by applying 1/8 inch above and 1/8 inch below the bust level. This increase must be applied to each piece that is involved in the grade.

Width Grading

Width grading refers to the amount of measurement added to a cross-body area, such as from shoulder point to shoulder point. A grade rule table is very helpful for determining these increments because they tend to be quite small comparison to other measurements.

Uneven Grading

Uneven grading is achieved when it is determined that the target customer is shaped somewhat differently than the standards in sizing chart. In these cases, the waist or hips may increase at a different rate than the bust as the design is graded to different sizes, producing a different shape of the finished garment to fit consumers who may be thicker through the hips and thighs or in the waist. The variations can be almost endless when combined with changes between styles and intended ease allotments.

Words and Phrases

increment [ˈinkrimənt] *n*. 档差
grade rule　放码规则，推档规则
bulk fabric　大货面料(大批量生产使用的面料)
full-size　原尺寸的，实际尺寸的
circumference grading
girth [gəːθ] *n*. 周长，围长
length grading　经向推档
width grading　纬向推档
uneven grading　不均匀推档

Chapter 5 SEWING / 服装工艺

LESSON 19 SEWING EQUIPMENTS / 缝纫设备

Sewing Tools	Hand Sewing Needles	Sized from coarse（#1）to very fine（#10）.
	Sewing Machine Needles	Sized fine to heavy. They should conform to the weight and type of fabric used.
	Pins	Satin steel pins recommended.
	Thimbles	An aid to efficient hand sewing.
	Beeswax	Used to coat thread for easier threading.
Cutting Tools	Bent-handle Dressmaker Shears	8 to 10 inches long, for cutting fabric.
	Thread Clippers	An efficient tool for clipping threads for both hand and machine sewing.
	Pinking Shears	9 or 10 inches long, used to give a zig-zag finish to raw edges of firmly woven fabrics.
	Seam Ripper	For ripping out unnecessary stitches and for opening machine-stitched buttonholes.
Measuring Tools	Tape Measure	Smooth-surfaced, clearly marked with centimeters as well as the familiar inches.
	Ruler	Clear plastic, marked with both centimeters and inches.
	Yard Stick	Available in 36- or 45- inch lengths, made out of metal or wood.
	Skirt Marker	For marking hem lengths accurately.
Making Tools	Tracing Wheel	Used to transfer pattern markings to fabric with tracing paper.
	Tracing Paper	Carbon-coated paper used in conjunction with the tracing wheel for transferring pattern markings to fabrics① .
	Graphite Paper	Colored coated graphite paper, available in art supply stores, used where carbon would not be visible.
	Tailor's Chalk	May be made of wax or stone. It is used to transfer marking to fabrics that will not accommodate carbon paper, and can also be used for marking adjustments in fittings and for hems.

续表

Sewing Machine Attachments	Cording Foot	Left-and right-handed. Use of either permits even, close stitching for cording and zippers[②].
	Piping Foot	For finishing bias binding strips.
	Gathering Foot	For even, permanent shirring.
	Roller Foot	Good for stitching on leather, synthetic suede, and vinyl.
	Invisible Zipper Foot	For stitching the invisible zipper.
	Narrow Hemmer Foot	For rolled machine hem stitching.
Miscellaneous Tools	Loop Turner	Used for turning bias strips to make "spaghetti" cord and narrow belts.
	Bodkin	Used for threading ribbon and elastic through a tunnel or casing.
	Dress Form	A professional dress form on an adjustable stand.
Pressing Equipment	Steam and Dry Iron	A heavy iron with adjustable temperature controls.
	Ironing Board	A well-padded shaped board, which stands firmly on the floor.
	Sleeve Board	A well-padded miniature board useful for pressing sleeves and small areas.
	Seam Roll	A padded roll for pressing hard-to-reach seams.
	Tailor's Ham	Used for pressing and molding curved areas.
	Tailor Board or Point Presser	An unpadded device helpful for pressing points.
	Needle or Velvet Board	Necessary for pressing velvet, velveteen, and napped and fur fabrics.
	Press Mitt	Used on the hand to press small areas without interfering with the rest of the garment.
	Press Cloth	Firmly woven drill cotton and wool cloths, necessary to protect fabrics from shine or scorching when pressing on the right side of the garment[③].
Others	The Lock Stitch Machine	Used for straight stitching.
	The Zigzag Machine	Used for appliqué lace, attach elastic, and provide a decorative finish for the edges of tricot knits.
	The Over-lock Machine	Combines straight and overcastting stitched and cuts and sews the fabric while stitching the seams[④].
	The Safety Stitch Over-lock Machine	Makes an additional row of stitching and uses 4 spools of thread.

Words and Phrases

coarse [kɔːs] *adj*. 粗的

fine [fain] *adj*. 细的

conform to　符合，遵守

thimble [ˈθimbl] *n*. （手工缝纫用的）顶针，针箍

beeswax [ˈbiːzwæks] *n*. 蜂蜡，蜂蜡色

bent-handle 弯把
thread clippers 纱剪刀
pinking shears 锯齿剪,花边剪
zig-zag 曲折线条,锯齿形,曲折形,人字形
firmly [ˈfəːmli] *adv*. 厚实地,坚实地,有身骨
seam ripper 拆线器
rip out 拆除,拆开
machine-stitched 机缝的
tape measure 软尺,带尺,卷尺
smooth-surfaced 光滑表面的
ruler [ˈruːlə] *n*. 直尺
yard stick 码尺,直尺
skirt marker 裁裙片样板
tracing wheel 插盘;点线轮,描样手轮
tracing paper 描图纸
graphite paper 石墨纸
tailor's chalk (裁缝用)划粉
wax [wæks] *n*. 蜡
attachment [əˈtætʃmənt] *n*. 附件
cording foot 嵌线压脚,滚边压脚
left-and right-handed 左右手的
cording [ˈkɔːdiŋ] *n*. 嵌线,包梗
piping foot 滚边压脚
gathering foot 碎褶压脚
shirring [ˈʃəːriŋ] *n*. 多层收皱,抽褶,平行皱缝
roller foot 双边固定轮压脚
suede [sweid] *n*. 人造麂皮,小山羊皮,起毛皮革,绒面革
vinyl [ˈvainil] *n*. 聚乙烯基织物
invisible zipper foot 暗拉链压脚
narrow hemmer foot 狭卷边压脚
miscellaneous [misiˈleinjəs] *adj*. 各种的,多方面的
loop turner 翻带器
spaghetti strap 细肩带
bodkin [ˈbɔdkin] *n*. 穿带用的钝针,粗长针;锥子
ribbon [ˈribən] *n*. 带,丝带,缎带,饰带
casing [ˈkeisiŋ] *n*. 抽带管;装嵌条
dress form 人体模型架,胸架
steam and dry iron 蒸汽电熨斗
ironing board 熨衣板
sleeve board 压袖板,烫袖板,袖子烫板;烫马
miniature [ˈminjətʃə] *adj*. 微型的,缩小的
seam roll 袖馒头,袖烫垫
hard-to-reach 难以达到的,难以触及的
tailor's ham 布馒头,馒头烫垫
point presser 马凳,烫凳,小烫台,小烫板
needle or velvet board (供绒毛织物整烫的)针毯烫垫
velvet [ˈveilvit] *n*. 天鹅绒
velveteen [ˈveiviˈtiːn] *n*. 棉绒,平绒
press mitt 手套式烫垫
press cloth 熨烫衬布,熨烫垫布
scorch [skɔːtʃ] *v*. 烧焦,枯萎
the lock stitch machine 锁式线迹缝纫机
the zigzag machine 锯齿形锁缝缝纫机,曲折缝缝纫机
tricot knits 经编织物
applique lace 贴饰花边
overcast [ˈəuvəkaːst] *adj*. 覆盖,遮盖,包边缝纫
the over-lock machine 包缝机,拷边机
the safety stitch over-lock machine 安全线迹包缝机
spool [spuːl] *n*. 线团

Notes

① Tracing Paper. Carbon-coated paper used in conjunction with the tracing wheel for transferring pattern markings to fabrics.
描图纸,用于在面料上复制纸样的复写纸,与点线轮配套使用。

② Cording Foot. Left-and right-handed. Use of either permits even, close stitching for cording and zippers.
滚边压脚，左右两侧均可使用，缝制效果均匀、紧密，用于包缝线迹和安装拉链。

③ Press Cloth. Firmly woven drill cotton and wool cloths, necessary to protect fabrics from shine or scorching when pressing on the right side of the garment.
熨烫垫布，质地较厚的斜纹棉和羊毛织物(制成)，防止面料正面熨烫时出现极光和烫焦。

④ The Over-lock Machine: Combines straight and overcastting stitched and cuts and sews the fabric while stitching the seams.
包缝机：用于裁切、缝制、包缝面料。

Discussion Questions

1. What is the distinguish between the hand tools and the machinery equipments?

EXTENSIVE READING

OPERATION OF SEWING MACHINES

Home sewers use either a straight-stitch or a zigzag sewing machine. In straight stitching, the needle moves up and down, producing a straight line of stitches; in zigzag stitching, the needle moves up and down and side to side, resulting in a zigzag line of stitching. The zigzag machine is equipped for decorative stitching, monogramming, over-casting, blind-stitching, sewing on buttons, making buttonholes, and mending.

Most modern sewing machines employ two separate threads to form a special type of stitch known as the lockstitch. The upper thread is led through an eye formed near the point of a needle. The under thread is carried on a bobbin and is linked or locked to the upper thread by means of a rotary or horizontal motion of the bobbin. In a typical machine employing a rotary bobbin, the sequence of operations is as follows. The needle carrying the upper thread moves downward through the material being sewed, and the thread is engaged above the eye of the needle by a hook on the rim of the bobbin. As the bobbin turns, the upper thread is pulled out to form a loop through which the under thread feeds.

The size of the loop is controlled by a tension device on the upper part of the machine. As the needle withdraws, the locked loop formed by the two threads is tightened by the pull of a lever take-up device to form a stitch. In a machine employing a horizontal bobbin held in a freely moving shuttle, the stitch formed is exactly the same. The shuttle moves through the loop of thread as the needle comes down, and then the shuttle returns to its original position as the needle moves up.

In addition to the large number of machines used for home sewing, thousands of different types of sewing machines are manufactured for industrial use. These include machines for the manufacture of hats, shoes, and hosiery, as well as for the sewing of garments. Modern machines, domestic and industrial, are often equipped with electronic controls such as microprocessors and other computerized devices to carry out automatic sequences of operations.

Words and Phrase

straight-stitch 直线线迹
stitch [stitʃ] *n.* 针脚,线迹;缝法,缝线
monogram [ˈmɔnəugræm] *n.* 字母组合,交织字母
over-casting 包边,锁边
blind-stitching 暗针,暗缝线迹
mending [ˈmendiŋ] *n.* 织补,缝补;修理
lockstitch [ˈlɔkstitʃ] *n.* 锁式线迹;锁缝
bobbin [ˈbɔbin] *n.* 筒管,筒子
rotary [ˈrəutəri] *adj.* 旋转的,轮转的
rim [rim] *n.* 边,缘;轮缘
lever [ˈliːvə] *n.* 杠杆
shuttle [ˈʃʌtl] *n.* (缝纫机的)滑梭
hosiery [ˈhəuziəri] *n.* 男袜,男用针织品
microprocessor [maikrəuˈprəusesə] *n.* 微处理机
computerize [kəmˈpjuːtəˌraiz] *v.* 给……装备计算机

LESSON 20 SEWING PROCESS / 缝制过程介绍

Sewing is mainly used to manufacture clothing and home furnishings. In fact, sewing is one of the important processes in apparel making. Most of such industrial sewing is done by industrial sewing machines. The cut pieces of a garment are generally tacked, or temporarily stitched at the initial stage. The complex parts of the machine then pierces thread through the layers of the cloth and interlocks the thread①.

Industrial Sewing

Although it seems to be a simple process, industrial sewing is quite a complex process involving many preparations and mathematical calculations for the perfect seam quality. Good quality sewing also depends on the sound technical knowledge that goes into pattern designing and making. Flat sheets of fabric having holes and slits into it can curve and fold in three-dimensional shapes in very complex ways that require a high level of skill and experience to manipulate into a smooth, wrinkle-free design②. Aligning the patterns printed or woven into the fabric also complicates the design process. Once a clothing designer, with the help of his technical knowledge, makes the initial specifications and markers, the fabric can then be cut using templates and sewn by manual laborers or sewing machine.

While handling the fabric and in the process of sewing, the cloth must be held stiff and unwrinkled. The seam quality is very sensitive to cloth tension that varies from time to time in the whole sewing process. These undesirable variations in the cloth tension affect the product quality. Therefore, there arises the need of strict control over the whole process. The work of sewing is focused on the handling of fabrics and guided them towards the sewing machines needle along the seam line. The attention is equally focused on the control of appropriate tensional force so as to maintain high quality seam③.

Pre-sewing Functions

Before the actual task of sewing begins, there are certain other tasks that have to be taken care of which can be termed as fabric handling functions-Ply separation; Placing the fabric on working table; Guiding the fabric towards sewing needle; and tension control of fabric during the sewing process④.

While ply separation, stacks of fabric plies are sequentially positioned with the help of some feeding apparatus in an unloading position. The uppermost ply in such a stack is individually and sequentially separated from the stack. In the process, an edge of the separated fabric is presented between the jaws of a gripping device. A proximity switch determines the spacing between the gripper jaws. When this spacing confirms that only one fabric ply is in place between the jaws, the single ply is transferred over onto the receiving end of a conveyor for further processing⑤. In aerodynamic technique, the uppermost ply of fabric is lifted by suction from the remainder of the stack.

When the fabric is placed on the working table, the tasks that are performed before the sewing process include-recognizing the fabric's shape, edges that will be sewn, planning of the sewing process and identification of the seam line.

Recognizing the Fabric's Shape

The appropriate tensional force depends on the fabric properties. So the fabrics have to be identified into categories like knitted fabrics, woven fabrics etc. depending on their physical properties. Various shapes of the fabrics such as, convex, non-convex, with straight or curve edges, also have to be considered and each of them require different handling strategies. In brief, the sewing methods done by automatic systems require classification of fabrics into various categories and certain preliminary scheme of the path that the fabric must follow so as to produce the required stitches⑥.

Fabric Edges to be Sewn

There are two basic types of stitches, one is that are for joining two parts of cloth together and the second one is done for decorative purposes. Sometimes, both types of stitching have to be done on some parts of cloth, for example, a pocket has to be joined on three sides with the apparel as well as it may be given some decorative stitches too. At what points and which type of stitching has to be done- all such information is stored digitally on automated devices through Computer-Aided Design (CAD) and accordingly sewn.

Planning of Sewing Process

Sequence of seams to be stitched is determined before the sewing starts. Which part will be joined first, what stitches will follow one another, etc. are decided. However, some stitches have to be necessarily done before or after another stitch. In the example above, the decorative stitches must be done first followed by the joining stitches. Embroidered patterns also follow the same sequence but sometimes in clothing items like hats, decorative stitches or embroidery is done after the production of hats and with the help of embroidery machine.

Identification of Seam Lines

Sewing process is performed on seam lines situated inside the fabric edges, some millimeters inside the fabric's outer line. For the straight lines, the seam line is found by transferring the outer lines inside the fabrics and the intersection of these lines makes the vertices of the seam line. Therefore, the seam line is parallel to the outer edge and the distance between the two has to be determined as it is different for different parts of the cloth. This distance is greater for trousers legs than for a shirt sleeves. Seam allowance is the area between the edge of the fabric and the line of stitching. It is usually 1.5cm away from the edge of the fabric. Seam allowance is usually 2.5cm or more for standard home dressmaking. Industry seam allowances vary but they are usually 0.6cm.

Sewing Fabrics

The sewing process consists of mainly three functions- guiding fabric towards needle; sewing of the fabric edge; and rotation around the needle. The fabric is guided along the sewing line with a certain speed that is in harmony with the speed of sewing machine. The orientation error is either manually monitored or if monitored automatically then error is fed to the machine controller so that the machine corrects the orientation of the fabrics. When one edge of seam line is sewed, the fabric is rotated around the needle till the next edge of the seam line coincides with the sewing line. The sewing process is thus repeated until all the edges of seam line planned for sewing, are sewed.

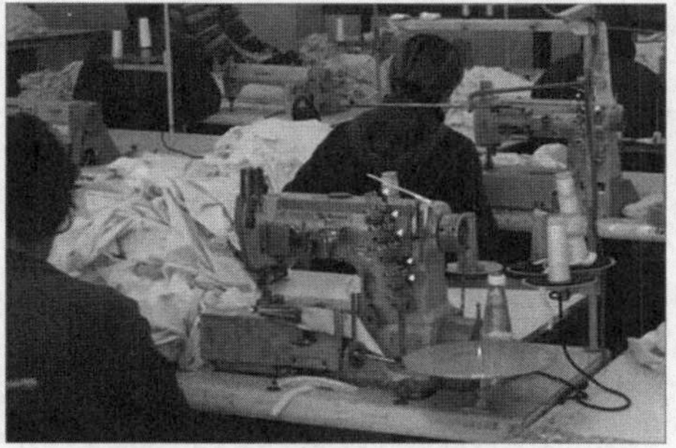

图 37 Choose a project that is equal to your sewing skill

The two main stitches that sewing machines make of which the others are derivatives are lockstitch and chain stitch.

- Back Tack.
- Backstitch-a sturdy hand stitch for seams and decoration.
- Basting Stitch (or tacking) -for reinforcement.
- Blanket Stitch.
- Blind Stitch (or hem stitch) -a type of slip stitch used for inconspicuous hems.
- Buttonhole stitch.
- Chain Stitch-hand or machine stitch for seams or decoration .

- Cross-Stitch-usually used for decoration, but may also be used for seams.
- Darning Stitch.
- Embroidery Stitch.
- Hemming Stitch.
- Lockstitch-machine stitch, also called straight stitch.
- Overhand Stitch.
- Overlock.
- Pad Stitch.
- Padding Stitch.
- Running Stitch-a hand stitch for seams and gathering.
- SailmakersStitch.
- Slip Stitch-a hand stitch for fastening two pieces of fabric together from the right side without the thread showing.
- Stretch Stitch.
- Topstitch.
- Whipstitch (or oversewing or overcast stitch) -for protecting edges.
- Zig-zag Stitch.

Words and Phrases

tack [tæk] *n*. 粗缝,假缝 *v*. 假缝;系、扎、扣、拴
pierce [piəs] *v*. 刺入;刺穿;穿透
wrinkle-free 不皱的
template [ˈtemplit] *n*. 模板
unwrinkle [ˌʌnˈriŋkl] *v*. 将(皱纹)弄平
sensitive [ˈsensitiv] *adj*. 敏感的,易受伤害的
arise [əˈraiz] *v*. 出现,发生
ply [plai] *n*. 织物层数
sequentially [sɪˈkwenʃəli] *adv*. 持续地,连续地
feeding [ˈfiːdiŋ] *n*. 推布,送料
apparatus [ˌæpəˈreitəs] *n*. 仪器,装置,器皿,用具
unloading [ˈʌnˈləudiŋ] *n*. 卸载
stack [stæk] *v*. 叠,堆,堆积于,把……堆积于
jaw [dʒɔː] *n*. 夹片,虎钳牙
grip [grip] *n*. 夹具,夹子 *v*. 夹紧,抓住,握住
proximity [prɔkˈsimiti] *n*. 接近,邻近
conveyor [kənˈveiər] *n*. 传播者,运送,运输设备
aerodynamic [ˌεərəudaiˈnæmik] *adj*. 空气动力学的,气体动力学的
uppermost [ˈʌpəˌməʊst] *adj*. 最上面的,最高的
suction [ˈsʌkʃən] *n*. 吸入,抽吸,抽气通风
remainder [riˈmeində] *n*. 剩余物,其余的人,余数
convex [ˈkɔnveks] *adj*. 凸的,凸面的
preliminary [priˈliminəri] *adj*. 预备的,开端的,初步的,开端
scheme [skiːm] *n*. 计划,方案
millimeter [ˈmilimiːtə] *n*. 毫米
vertex [ˈvɛːˌteks] *n*. 顶点
in harmony with 与……协调,与……一致
orientation [ˌɔːrienˈteiʃən] *n*. 方向,目标
back tack 倒回针
backstitch [ˈbækstitʃ] *n*. 回针,倒钩针
basting stitch 假缝线迹
blanket stitch 饰边缝线迹,锁缝线迹
blind stitch 暗针,暗缝线迹
buttonhole stitch 锁眼线迹

chain stitch 链式线迹
cross-stitch 十字线迹，人字线迹，十字形刺绣线迹
darning stitch 织补线迹
embroidery stitch 绣花线迹
hemming stitch 卷边线迹
lockstitch 锁式线迹，锁缝，锁针
overhand stitch 对接缝线迹
pad stitch 衬垫线迹，扎缚线迹；扎针，纳针
padding stitch 衬垫线缝，扎缚线缝；扎针，纳针
running stitch 初缝线迹，绗缝线迹
sailmaker [ˈseilˌmeikə] *n.* 用于帆布和厚皮革的针
slip stitch 短而松的暗缝线迹；挑针
stretch stitch 伸缩线迹
topstitch [ˈtɔpstitʃ] *n.* 正面线迹，面缝线迹，饰缝线迹
whipstitch [ˈhwipstitʃ] *n.* 搭缝；锁缝线迹
zig-zag stitch 曲折线条，锯齿形，曲折形，人字形

Notes

① The cut pieces of a garment are generally tacked, or temporarily stitched at the initial stage. The complex parts of the machine then pierces thread through the layers of the cloth and interlocks the thread.
服装的裁片一般先被固定，或者假缝起来，随后缝纫机上复杂的部件会牵引缝纫线穿过几层面料形成连结线迹。

② Good quality sewing also depends on the sound technical knowledge that goes into pattern designing and making. Flat sheets of fabric having holes and slits into it can curve and fold in three-dimensional shapes in very complex ways that require a high level of skill and experience to manipulate into a smooth, wrinkle-free design.
优质的缝纫还取决于对纸样绘制的正确操作。(例如)撕拉平整布料上的洞，布会抽缩并形成复杂的立体形状，这就要求高水平的技术和经验来恢复面料的平整。

③ The work of sewing is focused on the handling of fabrics and guided them towards the sewing machines needle along the seam line. The attention is equally focused on the control of appropriate tensional force so as to maintain high quality seam.
缝纫的工作重点就是控制面料，牵引面料沿着拼缝线在缝纫机上缝制。同时还需要保持最佳张力，才能确保高质量缝纫效果。

④ Before the actual task of sewing begins, there are certain other tasks that have to be taken care of which can be termed as fabric handling functions-Ply separation; Placing the fabric on working table; Guiding the fabric towards sewing needle; and tension control of fabric during the sewing process.
在开始缝纫前，需进行下列准备工作：分开整理多层裁剪的裁片；将对应的裁片放置在工作台上；正确放置面料方向；调整面料的松紧。

⑤ In the process, an edge of the separated fabric is presented between the jaws of a gripping device. A proximity switch determines the spacing between the gripper jaws. When this spacing confirms that only one fabric ply is in place between the jaws, the single ply is transferred over onto the receiving end of a conveyor for

further processing.
在此过程中，裁片的一边被放置在夹具的夹片之间，旁边的开关调节夹片之间的距离。当这个距离控制在只允许一层裁片时，这层裁片即被传送到一个运送器的接收端进行下一步操作。

⑥ In brief, the sewing methods done by automatic systems require classification of fabrics into various categories and certain preliminary scheme of the path that the fabric must follow so as to produce the required stitches.
总之，自动化缝纫设备的缝制方式要求面料（首先）分成若干个种类，并且必须遵守预先设计好的缝制顺序，才能得到理想的线迹。

Discussion Questions

1. Discuss the process of sewing a shirt.
2. What is the pre-sewing process?

EXTENSIVE READING

SPECIAL HANDICRAFT

Handicrafts are forms of art that are created by using your hands.

The Yarn Crafts

● Weaving

Weaving is the process of interlacing two sets of yarns to form fabric. One set of yarns, called the warp, is at right angles to another set, called the filling.

● Knitting

Knitting can also be done by hand or by machine. Knitting by hand requires knitting needles that are anywhere from 7 "to 14" (18—35cm) long. A single yarn is twisted and looped off of one needle and onto the other.

● Crochet

Crochet work looks similar to knitting, but it has a slightly lacis appearance. Crocheting is done with only one needle, called a crochet hook. The hook is used to pull the yarn through a loop, or a series of loops.

● Lace-Making

Lace-making is the most complicated of the knotting and looping arts. Many yarns are wound onto bobbins and then the bobbins are twisted and crossed around needles to create the patterns.

● Macramé

Macramé is the decorative art of tying knots. Macramé knots are often combined with beads threaded onto the hangers, jewelry, handbags, and other craft items.

● Braiding

Braiding is the process of overlapping and wrapping several strips of yarn, fabric, or leather around each other.

The Stitching Crafts

● Sewing

The most basic of the stitching crafts, is the art of using stitches to join pieces of fabric together.

● Embroidery

Embroidery is an art that includes many different types of decorative stitches used on fabric.

● Needlepoint

Needlepoint is the technique of forming stitches on a special open-weave fabric called canvas.

● Patchwork

Patchwork is the technique of cutting out small shapes of fabric and sewing them together to form larger shapes.

● Quilting

Quilting is the process of stitching together two layers of fabric with a soft material in between.

● Appliqué

Appliqué is the art of sewing one or more pieces of fabric to the top of a larger piece of fabric.

● Beadwork

Beadwork is a craft that was perfected by the North American Plains Indians. The beads are stitched to fabric to form intricate and colorful designs.

Dyeing and Printing

● Tie Dyeing

The fabric is tightly tied in certain places, then dipped into the dye. The dye will not penetrate in the spots where the fabric is tied.

● Batik Printing

Batik printing is created by first applying hot wax to the areas of the fabric that will not be dyed. The fabric is then dipped into a dye and left to dry.

● Block Printing

The block is covered with dye and pressed

图 38 Many people enjoy doing crafts such as macramé, needlepoint, counted cross-stitch, and braiding to produce things that are both decorative and useful

on the fabric so that the design is transferred from the black to the fabric.

- Silk Screen Printing

The screen is attached to a frame and a sealer is applied. This sealer covers all the areas of the screen that should not be printed. The screen is placed over the fabric, and the dye is applied to the screen. The dye seeps through the unsealed areas to create the design on the fabric.

- Painting

Painting on fabric is similar to painting on any other surface. The color can be applied with a brush, a pen, or a marker.

Words and Phrases

filling [ˈfiliŋ] *n*. 纬纱
crochet [ˈkrəuʃei] *n*. 钩针，钩针编制品；钩编花边
lacis [ˈleisis] *n*. 方网眼花边
crochet hook 钩针
lace-making 花边编织
knotting [ˈnɔtiŋ] *n*. 编结
macrame [məˈkraːmi] *n*. 流苏，花边
handbag [ˈhændbæg] *n*. 手袋
overlapping [ˈəuvəˈlæpiŋ] *n*. 重叠，交叠
needlepoint [ˈniːdlpɔint] *n*. 刺绣品；针绣
patchwork [ˈpætʃwəːk] *n*. 拼缝；拼缝品
quilting [ˈkwiltiŋ] *n*. 绗缝
beadwork [ˈbiːdwəːk] *n*. 珠绣
craft [kraːft] *n*. 手艺，手工艺，工艺
dip into 浸入
batik printing 蜡染
block printing 木板印染法
silk screen printing 丝网印
frame [freim] *n*. 结构，框架
sealer [ˈsiːlə] *n*. 保护层
seep through 渗透，渗过
painting [ˈpeintiŋ] *n*. 手绘

LESSON 21 SPECIFICATION / 工艺单

Specifications are exact criteria that must be met by a product or service. Apparel specifications define raw material requirements and how a garment is to be made to achieve the company's established quality standards①. In order for specifications to be meaningful, they are expressed in terms of numeric values. There can be a minimum or maximum acceptable value or a range of acceptable values called tolerances, which are allowable deviations from specified valued②. For example, in the case of fabric performance, the expectations for properties must meet a measurable maximum such as 2 percent shrinkage. With respect to dimensions, a tolerance is commonly used. For example, the length of the sleeve seam on a man's shirt could be 32 inches with an acceptable tolerance of 1/2 inch.

An apparel manufacturer must exercise care when using tolerances. For instance, if all of the dimensions of a garment are just within the upper tolerance limits, the entire garment may be too large③. In order to prevent this problem, a manufacturer can warn its suppliers that although the individual dimensions for a garment fall within allowable tolerances, the garment might be rejected if the overall measurements are not within tolerances. This warning must be stated in terms of a specific measurable quantity or percentage.

As in the case of standards, the writing of specifications must be exact, consistent, and absolutely clear to both supplier and purchaser. In order to achieve such precision, individuals known as specification or spec writers who are specially trained to perform this task may be hired. Spec writers must have superior knowledge of all dimensions of their product, be detail-oriented, exacting, and have the ability to write clearly and concisely④. Spec writers who are involved with fabrics require a mastery of fibers, spinning, knitting, weaving, color science, chemicals, and finishes.

A specification sheet is the document that communicates a garment's specifications both within and outside the company. The format for the specification sheet can come from a variety of sources.

A specification sheet is first used in the product development process when the prototype is created. At this time the designer needs to communicate to the pattern maker and sample maker exactly how the prototype should look.

A specification sheet would be used again to convey important information to the manufacturing facility. Normally this specification sheet is part of a package that would also include a production sample, a set of patterns, and possibly a production marker⑤. This specification sheet should include the following information:

- Style identification.
- Sketch or photograph.
- Sizes, the measurements for each size, and tolerances.
- Color-ways.
- Usage of fabric, trim, and findings.
- Construction details including seams and seam allowances, stitches, stitches per inch, placement of parts such as labels, pockets, etc.
- A suggested sequence of manufacturing operations.
- Information that is to be included on the care label.

Words and Phrases

define [di'fain] *v*. 定义,规定

established [is'tæbliʃt] *adj*. 建立的,固定的,既定的,确定的

meaningful ['miːniŋful] *adj*. 意味深长的,很有意义的

in terms of　用……的话,根据,按照

numeric [njuː'merik] *adj*. 数字的

minimum ['miniməm] *n*. 最小(量),最低额,最低点

deviation [ˌdiːvi'eiʃən] *n*. 偏离,偏向,偏差

expectation [ˌekspek'teiʃən] *n*. 期待,指望,展望

property ['prɔpəti] *n*. 财产,所有权,性质,属性

measurable ['meʒərəbl] *adj*. 可量的,可测量

supplier [sə'plaiə] *n*. 供应商
purchaser ['pəːtʃəsə] *n*. 买方,购买者
precision [pri'siʒən] *n*. 精密,精确
detail-oriented 针对细节,细心的
concisely [kən'saisli] *adv*. 简明地
mastery ['maːstəri] *n*. 精通,掌握
convey [kən'vei] *v*. 传达,运输,转让
format ['fɔːmæt] *n*. 开本,版式,形式,格式
maximum ['mæksiməm] *n*. 极点,最大量,极大

Notes

① Specifications are exact criteria that must be met by a product or service. Apparel specifications define raw material requirements and how a garment is to be made to achieve the company's established quality standards.
工艺单是产品或服务需要严格遵守的标准。服装工艺单规定了原材料的要求、服装的制作过程,目的是为了达到公司设定的质量标准。

② There can be a minimum or maximum acceptable value or a range of acceptable values called tolerances, which are allowable deviations from specified valued.
最大或最小的可接受数值或一个可接受的数值范围称为公差,允许与标准尺寸有一定的偏差。

③ An apparel manufacturer must exercise care when using tolerances. For instance, if all of the dimensions of a garment are just within the upper tolerance limits, the entire garment may be too large.
一个服装制造商必须小心使用公差。例如一件服装的所有尺寸均处于上公差范围,那么整个服装就可能太大。

④ In order to achieve such precision, individuals known as specification or spec writers who are specially trained to perform this task may be hired. Spec writers must have superior knowledge of all dimensions of their product, be detail-oriented, exacting, and have the ability to write clearly and concisely.
为了达到这个精确度,只有受过专门训练的人才会被雇用编写工艺单。工艺单编写人必须对产品的所有尺寸有充分的了解:细心、严格、条理清晰、准确。

⑤ A specification sheet would be used again to convey important information to the manufacturing facility. Normally this specification sheet is part of a package that would also include a production sample, a set of patterns, and possibly a production marker.
规格表还将主要的信息传递回制造环节。通常规格表是一个文件包的一部分,文件包还包括产品样板、一套纸样、可能还有生产的排料图。

Discussion Questions

1. What differences exist between the size categories for body types in women's wear and what might these differences mean to garment styling?

EXTENSIVE READING

SPECIFICATION SHEET

Exact Knitting Mills

UPSPACE APPAREL

Measurement Specifications

Created By:	Mike Jackson 2/12/09 12:30 PM
Last Modified By:	Mike Jackson 2/12/09 9:30 AM

Style	T990 Cap Sleeve 4-Button Tee	**Description**	Sample BP 111		
Sample Dimensions	XS	**Color**	Assorted	**Version**	1
Season	All Year	**Class**	Junior Tops	**Approved by**	Bob Alex
Customer	Clair Casual	**Status**	Development	**Style copied**	New

POM Item	Description	Low TOL	High TOL	Measure Base	XS	S	M	L	XL
T100	Front Body Length Fr HPS	0.00	1.00	inches	23.50	24.00	24.50	25.25	26.00
T200	Chest Width 1# Below Armhole	−0.50	0.50	inches	10.00	11.00	12.00	13.25	14.50
T212	Sweep	−0.75	0.75	inches	11.75	12.75	13.75	14.75	15.75
T231	Waist	−0.50	0.50	inches	8.50	9.50	10.50	11.75	13.00
T186	SHLDR Width	−0.25	0.25	inches	3.125	3.25	3.50	3.625	3.75
T187	SHLDR Width Pt. To Pt-Straight	−0.375	0.375	inches	12.50	13.00	13.50	14.00	14.50
T704	Neck Width Edge To Edge	−0.125	0.125	inches	7.50	7.75	8.00	8.125	8.25
T230	Waist Position Fr HPS	−0.001	0.001	inches	13.75	14.125	14.50	15.00	15.50
T222	Across Frt.6# Fr HPS(Raglan)	−0.50	0.50	inches	10.75	11.25	11.75	12.375	13.00
T223	Across Bk.6# Fr HPS(Raglan)	−0.50	0.50	inches	11.50	12.00	12.50	13.125	13.75
T301	Sleeve Length Fr CB-Short	−0.25	0.25	inches	9.75	10.00	10.25	10.625	11.00
T311	AH Depth-Raglan	−0.25	0.25	inches	7.875	8.125	8.375	8.75	9.25
T335	Sleeve Opening-Short Sleeve	−0.25	0.25	inches	4.125	4.375	4.75	5.25	5.75
T713	Front Neck Drop From HPS To Edge-AA	−0.125	0.125	inches	7.75	7.875	8.00	8.125	8.25
T714	Back Neck Drop From HPS To Edge	−0.125	0.125	inches	0.25	0.375	0.50	0.625	0.75
T180	Underarm Sleeve Length	−0.25	0.25	inches	0.625	0.625	0.625	0.625	0.625

An example of a specification sheet that defines the graded measurements for a woman's shirt.

Words and Phrases

sample dimensions 样衣尺寸
junior tops 少年上装
assorted [ə'sɔːtid] *adj*. 分类的，配色的
Fr. from 的缩写
HPS (hips) *n*. 臀围；臀部
chest width 胸宽
sweep [swiːp] *n*. 下摆围
SHLDR. shoulder 的缩写
SHLDR Width 肩宽
Pt. point 的缩写
Frt. front 的缩写
raglan ['ræglən] *n*. 连肩，包肩
Bk. back 的缩写
sleeve opening 袖口
neck drop 领口深
underarm ['ʌndəraːm] *n*. 腋下
measure base 测量单位
TOL. tolerance ['tɔlərəns]的缩写 *n*. 公差；容限；容许数；宽松度
Low TOL 下误差
High TOL 上误差

LESSON 22 EVALUATION / 质量评估

Quality can be evaluated in terms of the characteristics of the completed product, as well as by the characteristics of the techniques used in its production.

A garment is theoretically evaluated from the time a pattern is selected to the last fitting and final pressing. A pattern should be selected to suit the individual's personality as well as his or her figure, and, if such a pattern cannot be found commercially, several parts can be combined, or one can design his or her own.

Many judges in the fashion field are not interested in seeing the inside construction of a garment, as they firmly believe that any garment will show from the outside appearance any imperfections that could be found in the construction from the interior of the garment①. For instance, a seam which has not been sewn accurately from the interior of the garment will show up more distinctly on the exterior of the garment when the garment is modeled.

In evaluating clothing construction, the technique used in constructing the garment can become very important. It must suit the fabric's fiber content and type, the style of the garment, and the intended use of the garment②.

GENERAL EVALUATION FOR CLOTHING CONSTRUCTION SCORE SHEET

Total Points Possible		Points	Fair	Good	Excellent	Comments
Selection of Fabric for Project	Suited to the individual					
	Suited to the pattern					
	Suited to skill					
	Thread —Correct type —Correct color					
	Suited to fashion fabric					
	Interfacing and underlining —Weight —Location					
	Lining —Color —Weight —Suited to garment					
Neatness and Accuracy of Workmanship	Machine stitching —Correct stitch length —Balanced tension					
	Seam finish —Appropriate type for fabric —Appropriate location for fabric					
	Hand stitching —Appropriate stitch —Even in length —Appropriate length					
	Pressing —Free from shine —Free from press marks —Free from wrinkles —Well-pressed seams					
	Accuracy of seam construction Straight seams —Matched accurately —Stitched straight Curved seams —Matched accurately —Stitched smoothly					

续表

<table>
<tr><td rowspan="2"></td><td>Corresponding sides of garment piece similar in length and width
—Collar points
—Length of bodice seams
—Length of skirt seams
—Corresponding darts
—Shoulder seams
—Buttonholes
—Decorative detail</td><td></td><td></td><td></td><td></td><td></td></tr>
<tr><td>Accuracy in matching sections of garment
—Corresponding darts
—Plaids
—Stripes
—Motifs
—Floral</td><td></td><td></td><td></td><td></td><td></td></tr>
<tr><td rowspan="7">General Appearance</td><td>Darts
—Tapered correctly
—Thread secured
—Pressed in correct position</td><td></td><td></td><td></td><td></td><td></td></tr>
<tr><td>Facings
—Seams clipped or notched
—Seams graded
—Seams understitched
—Reinforced with interfacing</td><td></td><td></td><td></td><td></td><td></td></tr>
<tr><td>Waistline treatment
—Stay correctly placed
—Stay correctly stitched
—Seams pressed in correct direction</td><td></td><td></td><td></td><td></td><td></td></tr>
<tr><td>Gathers
—Fullness evenly distributed
—Free from tucks or pleats</td><td></td><td></td><td></td><td></td><td></td></tr>
<tr><td>Pleats
—Fullness evenly distributed
—Even in width</td><td></td><td></td><td></td><td></td><td></td></tr>
<tr><td>Buttonholes
—Correct placement
—Correct size
—Lips even and in standard size</td><td></td><td></td><td></td><td></td><td></td></tr>
<tr><td>Pockets
—Placement
—Welt even in size
—Correct size
—Fit of welt into pocket
—Patch placement
—Type of construction used</td><td></td><td></td><td></td><td></td><td></td></tr>
</table>

续表

	Zippers —Top stitching even in width —Underlap hidden —Slot zipper seam meets in center —Correct length for use —Correct type for placement —Correct color for fabric					
	Sleeves —Set-in sleeve eased in smoothly —Cut-in-one sleeve reinforced or gussets					
	Collars and cuffs —Seams clipped or notched —Seams graded —Seams understitched or under collar cut on bias					
	Hems —Even in width —Appropriate width —Correct finish for fabric —Correct finish for style of garment —Use of seam tape and appropriate color for fabric —Appearance from the right side of garment					
	Belt —Suitable width —Neatly finished					
	Fasteners —Correctly placed —Appropriate size —Appropriate color —Securely fastened —Covered snaps —Buttons with shanks of appropriate length					
	Extras —Hand pricking —Self-trim					

Words and Phrases

theoretically [θiə'retikli] *adv*. 理论上，理论地

figure ['figə] *n*. 外形，形状，体形

imperfection [ˌimpə'fekʃən] *n*. 不完整性，不足；缺点

distinctly [dis'tiŋktli] *adv*. 显然地，明显，

清楚地
intended [in'tendid] *adj*. 有意的，故意的
underlining [ˌʌndə'lainiŋ] *n*. 衬里
workmanship ['wəːkmənʃip] *n*. 手工，工艺
clip [klip] *v*. 剪去，修剪，剪短
understitch [ˌʌndə'stitʃ] *v*. 暗缝
underlap [ˌʌndə'læp] *n*. 里襟
cut-in-one 连裁
undercollar [ˌʌndə'kɔlə] 领里，领下片
shank [ʃæŋk] *n*. 杆，身，柄部
self-trimming 同料边饰

Notes

① Many judges in the fashion field are not interested in seeing the inside construction of a garment, as they firmly believe that any garment will show from the outside appearance any imperfections that could be found in the construction from the interior of the garment.
时装领域的许多评判不在于观察服装的内部结构，因为他们笃信任何服装外观上的不足均能在内部结构上找到原因。

② In evaluating clothing construction, the technique used in constructing the garment can become very important. It must suit the fabric's fiber content and type, the style of the garment, and the intended use of the garment.
在评估服装的结构方面，服装工艺制作的方法很重要，它必须适合面料纤维的成分和种类、服装的款式以及服装的用途。

Discussion Questions

1. Why and how to evaluate the quality of a garment?
2. Which characteristics of the garment is the key point in the evaluation?

EXTENSIVE READING

STEPS IN THE PRODUCTION PROCESS

The first step in producing a new product is to estimate the production run, or the quantity that is required to meet the needs of the retail stores. At this point, a costing sheet and specifications package are used to estimate the cost of each individual product. Once the factory and the designer/retailer agree on a price, production can begin.

First, the factory must make a pattern. Then, in a process called pattern grading, the factory makes the pattern to different sizes. Grading can be done either via computer or by hand, but today, is normally done by computer. Once the pattern is complete, and the factory receives approval from the designer, cutting of the pattern can begin.

The factory will then cut a few samples from the agreed-upon fabric. These samples will then be sent to the designer for approval. If time is a concern, the factory may not send the actual completed garment but photos of it instead. After the factory has received approval for the finished product, it will produce the number of garments requested, usually using an assembly line that runs the full length of the factory floor. Garments are normally produced in sections. A worker would sew one or two specific sections of the garment, then pass it onto the next person to complete next section. The final processes include sewing the trimmings, buttons, and other details on the garment.

The factory, if contracted to do so, will tag the garments with the retailer's tag, sku number, and price. At this point, the garments will be shipped to the appropriate distribution center.

The process for reordering garments is much easier. If an item has sold well in a store, a designer or buyer may only have to call or e-mail the factory and ask for additional quantities of the garment.

To produce goods overseas, the designer would identify the origin of goods and locate a manufacturer. Then, the factory would see samples, at which time it would provide prices. Payment terms would be agreed upon, then the designer would apply for necessary licenses and arrange for the clearing of the goods through customs.

Steps in the production process:

Estimate production run/this involves utilization of specifications package and costing sheets.

Making of the pattern/here the pattern is graded in the correct sizes.

Sewing samples/using the pattern, samples are sewn and sent to the designers for approval.

Mass production of approved garment/the factory will then cut and sew, then add trimmings, such as zippers, to the garment.

Tagging and shipping/after they are tagged, the garments are ready to be shipped to the retailer's distribution center.

Words and Phrases

via [ˈvaiə] *prep*. 通过
approval [əˈpruːvəl] *n*. 批准,认可,确认
agreed-upon 达成协议的
assembly line 装配线,生产流水线
tag [tæg] *v*. 标上……标签,标签
sku number 单品数量
reorder [ˈriːˈɔːdə] *v*. 翻单,追加订货
payment term 支付条款
license [ˈlaisəns] *n*. 许可证

LESSON 23 LABELS AND HANGTAGS / 标签和挂牌

Every garment must now have one or more labels that give the consumer specific information. This information is mandatory, and required by law[①]. These labels should be attached to the garment in places where they are easy to find. Usually, this is at the center back of a garment-at the neck of a shirt, sweater, or blouse, and at the waist of pants and skirts. Sometimes you will find them in the inside lower front seam of a jacket or in the side seams of lingerie.

These labels can be glued, sewn, printed, or stamped on the fabric. They can even be attached to the outside of the garment as long as they are put on in such a way that they will not come off until you, the purchaser, decide to take them off.

However, most labels remains permanently attached to the garment. If the garment will remain in a package until after it is sold, the fiber content label can be affixed to the package only.

According to the TFPI[③] Act, there are five pieces of mandatory information that must appear on labels in the garment:

1. Any fiber that makes up 5 percent or more of the garment by weight must be listed.

2. Percentage of Fiber Content by Weight.

The fibers must be listed in descending order by percentage. This means that the fiber present in the greatest amount is listed first. Fibers present in an amount less than 5 percent are listed last, as "Other Fibers". If a garment is made of only one fiber type, the label must say so. It can say "100% Cotton" or "All Cotton"[②].

3. Identification of the Manufacturer.

This label identifies who is responsible for the product. Either the name of the manufacturer or the store, the registration identification number, or the trademark name for the product is written on the label. A trademark is a symbol, design, word, or letter that is used by manufacturers or retailers to distinguish their products from those of the competition. Trademarks are registered and protected by law so that no one can use someone else's trademark.

图 39 It is useful to know where hangtags and labels are usually found, since there may be more than one of each type on a garment

4. Country of Origin.

The label must state where the garment was manufactured, such as "Made in India" or "Made in Hong Kong".

5. Care Requirements.

Information on how to take care of the garment is regulated by the Care Labeling Rule, issued by the Federal Trade Commission.

Hangtags are labels that literally hang from the garment. They can be attached with a string, a thread, a strip of plastic, or a safety pin. Before you wear the garment, you remove these labels. Hangtags may repeat some of the information that appears on labels within the garment. However, the information that is present on the hangtags is voluntary, given freely and not regulated by law.

Manufacturers' hangtags may include the following information:

- The trademark or brand name of the manufacturer. This can take the form of the logo, or symbol, for the product or its manufacturer.
- The trademarks, brand names, or logos for the fibers.
- Information about the construction of the fabric. This information may tell whether the fabric is a stretch fabric or a knit fabric.
- Information concerning any warranty or guarantee that applies to the garment or its fabric.

Words and Phrases

hangtag [ˈhæŋtæg] *n*. 使用说明,飘带式商品标签,吊牌

mandatory [ˈmændətəri] *adj*. 受委托的,强迫的,强制的

blouse [blauz] *n*. 衫;上衣;罩衫;宽大短大衣

pants [pænts] *n*. 长裤,便裤

glue [gluː] *v*. 胶合

come off 发生,表现

affix [əˈfiks] *v*. 粘贴

descending [diˈsendiŋ] *adj*. 递降的

registration [ˌredʒisˈtreiʃən] *n*. 登记

trademark [ˈtreidmaːk] *n*. 商标

competition [ˌkəmpiˈtiʃən] *n*. 竞争,竞争者

country of origin 原产国,原产地,出产地

care requirement 使用须知

literally [ˈlitərəli] *adv*. 按照字面上地

string [striŋ] *n*. 细绳,绳索

voluntary [ˈvələntəri] *adj*. 自发的,自愿的

logo [ˈləgəu] *n*. 图形,商标,图样

warranty [ˈwərənti] *n*. 保证(书),根据理由,授权

guarantee [ˌgærənˈtiː] *v*. 保证,担保;*n*. 保证书,保证人

Notes

① Every garment must now have one or more labels that give the consumer specific information. This information is mandatory, and required by law.

现在每件服装必须有一个或多个标签,给消费者详细的信息。这些信息是法定的,必须执行。

② The fibers must be listed in descending order by percentage. This means that the fiber present in the greatest amount is listed first. Fibers present in an amount less than 5 percent are listed last, as "Other Fibers". If a garment is made of only one fiber type, the label must say so. It can say "100% Cotton" or "All Cotton".

必须按降序列出纤维百分比。首先是主要成分的纤维，最后是成分不足5%的“其他纤维”。如果服装只使用了一种纤维，标签必须列出，它可以标成“100%棉”或“纯棉”。

③ TFPI：The Textiles Fibers Product Identification Act.
纺织品成分标签法（美国）

Discussion Questions

1. What is the function of the label?
2. Watching a label and talk about the main contents on it with your classmates.

EXTENSIVE READING

LABELS：FINISHING PROCESS

Application

- Sewn. All woven labels and most printed labels are sewn with a sewing machine onto a garment or other product. As a rule, there should be a 1/16“to 3/16” sew space allowance on the edge(s) of the label, but this allowance varies depending on the cut of the label.
- Adhesive. Printed labels only can be applied flat on a garment or other product using an adhesive backing that is either activated through heat transfer or features a self-adhesive sticker back. Once the protective paper is peeled off, the label is ready for application. In both these processes, the label must be finished as a “cut & seal” label with no folds.

Woven Label Finishing Options

- Folding . Woven labels can be finished in a variety of cuts and folds, determined by where they are being attached on the garment and the styling desired. Most of these folds can also be applied to printed labels.
- End Fold. When the left and right sides of the label are folded back (tucked under). The label is then sewn onto the garment at the folds, providing an attractive folded edge. This is one of the most popular folds, appearing on countless garments. When it's the top and bottom that are folded back, the label is called a Top & Bottom Fold.
- Centerfold. Also known as a Loop Label, because the label is folded in half and then the two cut ends are both sewn together into the garment, often at the neckline or waistband. The Centerfold is ideal for care/content labels, because in effect it creates a two-sided label with ample room for text. Many garments will feature the brand identity on the front and the care/content information on the back.
- Cut & Seal. This label is cut straight on all four sides, with no folds. This the best

choice when the label will be sewn on all four sides, including outside pocket patches but also inside the neck or waist of many tops and bottoms.

- Mitre Fold. Both ends of this label are folded up diagonally, then sewn into the garment, usually at the neckline of tops. It creates a loose "loop" associated with hanging on closet hooks.
- Die Cut. Die cutting allows the creation of almost any shape imaginable, and is the right choice for any label that's not square or rectangular. The process is more expensive than other options, because each label is cut by hand. A label can even feature a die cut "window" to create a "see-through" effect.
- Continuous. Also known as Woven Tape, this is the correct specification when you need flexibility at your sewing facility, with the choice of cutting on-site for individual labels or running a continuous "tape" down the side of a pant leg.
- Cutting. Special cutting methods have been developed to avoid the scratchy feeling that the consumer may feel when a label rubs against his or her skin.
- Hot Knife. By far the most common method, the woven label is cut with a heated blade, which melts the polyester fibers at the edge, reducing rough residue that might be scratchy.
- Ultra-Sonic. Slower and more expensive than Hot Knife, the Ultra-Sonic cut uses high-frequency sound-wave technology, achieving a supremely smooth edge.
- Starching. Many of our customers request their labels be starched to help make the label more rigid and easier to handle in the sewing facility. This is particularly helpful for smaller labels, and washes out after the first wash by the consumer, with no permanent effect on the label.
- Filled Woven Labels. Woven labels can be stuffed with yarn and heat-sealed to create a "puffed" effect, ideal for zipper pulls or special effects on garments.

Printed Label Finishing Options

- Folding. Printed labels can be cut and folded much like woven labels. Folds include End Fold, Centerfold, Cut & Seal and Mitre Fold. Almost all printed labels are square or rectangular. If you need a non-standard die cut shape, a woven label is the better choice.
- Cutting. Printed labels created on fabric can be cut like woven labels using Hot Knife or Ultra-Sonic cutting Printed labels created on Tyvek or other synthetic materials are usually chop-cut, adding to their affordability.

Words and Phrases

adhesive [əd'hiːsiv] *n*. 粘合剂

sticker ['stikə] *n*. 缝纫机;缝纫工

peel off　去皮,剥离

seal [siːl] *v*. 封印,密封,隔离;图记;记号

countless ['kaʊntlis] *adj*. 无数的,数不尽的

top & bottom (织物)上下端

centerfold [ˈsentəˌfəʊld] *n*. 中间折叠

ample [ˈæmpl] *adj*. 容量大的

mitre [ˈmaitə] *n*. 45°折角拼缝,斜拼接,斜接缝

die [dai] *n*. 刀模

on-site 现场的

scratchy [ˈskrætʃiː] *adj*. 使人发痒的,扎人的

hot knife 热封刀,热封钳

ultra-sonic 超音波

sound-wave 声波

starch [staːtʃ] *v*. 浆洗

heat-sealed 热封的

Tyvek [ˈtaivik] *n*. 高密度聚乙烯合成纸

chop [tʃɔp] *n*. 商标

Chapter 6 FASHION MARKETING / 服装营销与市场

LESSON 24 THE SRTUCTURE OF THE FASHION MARKET / 时装市场的组成

Apart from technology, the reason why fashion is now available to the masses is that there are several levels at which clothing functions:

- Haute Couture House.
- Designer Wear.
- Street Fashion/Mass Market.

Haute Couture House

They are the major fashion houses of the world, run by recognized, international famous designers. They show their collections at least twice a year and sell individual garment at very high cost. For these designers, the catwalk shows are essentially a publicity exercise for their garments, perfumes and accessories.

In recent year, the haute couturier have, along with established designers, tend to move towards greater brand differentiation to capitalize on their names and some have also decentralized their manufacturing operations to cut costs①.

Some manufacturers now produce and distribute designer collections enabling haute-couture or designer names to be made available to a large market at more accessible prices through ready-to-wear ranges. The announcement by the French government in 1922 that was planning to encourage new designers into haute couture indicated the fact that the couture market is in decline②.

Younger customers are being tempted away from the idea of luxury for its own sake and are now demanding clothes by the newer designers that are indisputably contemporary in their direction and approach.

Designer Wear

Designer wear is shown at the "Prêt-a-Porter". The move into ready-to-wear clothing by designer meant that they could offer their stylish designs and high quality to a wider audience, they are to be found in designer's shops, independent stores and some of the more exclusive department stores. Designs are not limited, but are still produced in limited numbers. A ready-to-wear designer has the same aim as the haute couture designer in creating flattering, attractive fashionable garments③. A ready-to-wear fashions are usually less innovative and imitate the fashions at haute couture level. They may create a similar version of a particular outfit that was an outstanding seller a year back.

Mass Market or Street Fashion

It is a market area in which most people buy their clothes. New fashions can be seen in the high street stores very quickly. This is an area of the market that is undergoing many changes.

图 40 Street fashion always be casual

The three tier view of market is perhaps over simplistic as there are many strata and price levels between the ones mentioned④. Many customers do not stick to any one level when buying their clothes. The more affluent will buy several haute couture outfits but turn to designer wear for everyday. Women who mostly buy designer ready-to-wear may occasionally splash out on a couture dress for a very special occasion. Those who generally only buy mass marketing clothing may still buy designer wear occasionally if only from the discounted retail outlet.

Second-hand clothing-in some high street shopping centers charity shops seems to be almost as common as new clothing shops⑤. Because of the huge increase in the number of these, the second-hand clothing outlets can be explained in many ways.

图 41 A small company invites a young model to try on some new styles

Words and Phrases

designer wear 设计师(标名)服装

perfume [ˈpəːfjuːm] *n.* 香料,香水

differentiation [ˌdifəˌrenʃiˈeiʃən] *n.* 区别,分化,变异

capitalize [kəˈpitəlaiz] *v.* 使资本化,估计……的价值

announcement [əˈnaunsmənt] *n.* 公告,发表,告知

decentralize [diː'sentrəˌlaiz] *v*. 分散
indicate ['indikeit] *v*. 指出,表明,表现出
tempt away 诱惑,引诱,拐走
luxury ['lʌkʃəri] *n*. 奢侈,豪华,奢侈品
sake [seik] *n*. 目的,缘故,理由
indisputably [ˌindi'spjuːtəbli] *adv*. 无可争辩,无可置疑
Prêt-a-Porter 现成的服装,高级女装成衣
imitate ['imiteit] *v*. 模仿
tier [tiə] *n*. 等级
simplistic [sim'plistik] *adj*. 简单化的
stratum ['straːtəm] *n*. 阶层
affluent ['æfluənt] *n*. 富人
splash out 挥霍钱财
second-hand 用旧的,二手的
charity ['tʃæriti] *n*. 慈善机构

Notes

① In recent year, the haute couturier have, along with designers, tend to move towards greater brand differentiation to capitalize on their names and some have also decentralized their manufacturing operations to cut costs.
近些年来,高级订制师和知名服装设计师均开始出现细分品牌和资本化品牌的迹象,有些品牌已经开始分散制作加工来降低成本。

② Some manufacturers now produce and distribute designer collections enabling haute-couture or designer names to be made to a large market at more accessible prices through ready-to-wear ranges. The announcement by the French government in 1922 that was planning to encourage new designers into haute couture indicated the fact that the couture market is in decline.
一些制造商筹备和推广设计师发布会,这使得高级订制或设计师品牌服装能以成衣的形式在巨大的低价市场销售。1922 年法国政府颁布的鼓励新设计师进入高级订制业的声明,就反映出高级订制正在衰退。

③ A ready-to-wear designer has the same aim as the haute couture designer in creating flattering, attractive fashionable garments.
高级时装成衣设计师和高级订制设计师有着共同的目标,就是设计优美并吸引人的时尚款式。

④ The three tier view of market is perhaps over simplistic as there are many strata and price levels between the ones mentioned.
市场的三等级观的划分也许过于简单了,因为在这些市场中还可以细分成许多档次和价位。

⑤ Second-hand clothing in some high street shopping centers charity shops seems to be almost as common as new clothing shops.
二手服装在高级社区中心的慈善商店出现的频率和在新品店差不多。

Discussion Questions

1. How much is the customer willing to pay the most fashion style in this season?
2. What blends of fibers do consumers like best in your city?
3. What fabric finishes are most desired?

EXTENSIVE READING

THE MASS PRODUCTION PROCESS

The clothing industry is as mechanized as any in the industrialized world and the number of workers that employs compass with giants such as the motor and other product manufacturing industries. The route from design to finished garment is complicated and in many ways mimics the processes that the original dress-maker or tailor used in the past. These processes have been broken down into a sequence: the pattern, the sizing, the lay plan, the cutting out, the sewing together and the finishing.

Before the pattern is made the fabric is tested for abrasion, pillage, shrinkage and so on. The pattern has to ensure that not only the desired silhouette, details, fit, hang and fall match the designer's sketch but also that built into this pattern are instructions on how the garment is to be assembled. This is achieved by a series of coded marks that denote the width and form of the seam allowances, notches to indicate where the seams are to be, drill holes to show darts and pocket placing. The correct sizing is achieved by a process called grading which is a technique that uses the master pattern to develop a series of increasing or decreasing sizes for each design which can be anything from size 8 to 22 and even higher. Nowadays grading is usually done by computer and relies on complicated mathematical calculations. The introduction of computer-aided grading has speeded up this process ten-fold, but the computer is only a tool, it is the grader who is the irreplaceable expert in the grading process.

The next stage is the lay plan which is a method of placing pattern pieces on to the cloth economically, not unlike a jigsaw puzzle. This placing is affected by the particular characteristics of the fabric: grain lines, checks and stripes, pile or one-way fabrics. The cutting out is either manual or automated. If manual, the cloth is laid up either by two people or by one person operating a cloth-lay spreading machine. The cloth is laid up in layers with tissue paper or its equivalent between each roll of fabric to distinguish dye batch and color variation. A minimum of 5 layers to a maximum of 100 can be laid up in this way, depending on the thickness of the cloth and accuracy required. A manual lay is cut using either a band knife or a straight knife.

Multi-layered automated cutting is now carries out on a specially designed vacuum-com-pressed table where 25 layers can be laid and cut at once. The cutting is done with a mechanical band knife and is now mostly computer-aided and entirely automatic. The cut pieces are bundled together and docketed.

Assembly is where the garment is sewn together. This assemble is itself broken down into component parts; the collar, pockets, sleeves and cuffs for example will be assembled separately and then attached to the garment body or shell which will itself have been assembled separately in advance. Each of these processes is carried out by a different operative who works on a mechanized assembly line, now usually also computer-aided so that, for example, the collar is dropped off at the machinist with the most collar expertise, the over-locking at the over-locker and so on. Each design will have a different assembly plan worked out in advance on the prototype which will usually have been assembled by one expert sample machinist with the help of a garment technologist.

The finishing, until a short time ago done by hand, is now mechanized, with the technical innovations in blind hemmers superseding the hand finishers. In some manufacturing companies, pressing takes place during assembly and is called "under pressing" but the usual practice is to use one of the sophisticated pressing units to press the finished garment and possibly to blow hot air through it. This is then passed by the quality controller and packed ready for distribution.

Words and Phrases

industrialize [in'dʌstriəlaiz] *v*. 使工业化，实现工业化
compass ['kʌmpəs] *v*. 达成,完成
giant ['dʒaiənt] *adj*. 庞大的,巨大的
mimic ['mimik] *n*. 模仿,仿制品
lay plan 铺料
pillage ['pilidʒ] *v*. 起球 *n*. 起球
coded mark 编号,记号
denote [di'nəut] *v*. 指示,表示
irreplaceable [ˌiri'pleisəbl] *adj*. 不能替代的
economically [iːkə'nɔmikəli] *adv*. 节约地,不浪费地,节省地
master pattern 基本纸样,母板
jigsaw puzzle 拼图玩具
one-way 一顺的,单向的
manual ['mænjuəl] *adj*. 手动的,手工的
automate ['ɔːtəmeit] *v*. 自动作业,使自动化
cloth-lay 铺料
tissue paper 薄纸
equivalent [i'kwivələnt] *n*. 同等物,相等物
batch [bætʃ] *n*. 批,群,组;成批生产
color variation 色差
band knife 带刀
straight knife 直条裁剪刀
multi-layer 多层
vacuum-com-pressed 吸风烫台
bundle ['bʌndl] *n*. 捆,束,包 *v*. 捆,扎
docket ['dɔkit] *n*. 标签,签条,标志
assembly line 装配线,生产线
machinist [mə'ʃiːnist] *n*. 机械师,机械修理工
blind hemmer 暗卷边机
supersede [ˌsjuːpə'siːd] *v*. 代替,接替,更替
under pressing 缝前熨烫,半成品熨烫

LESSON 25 MERCHANDISING PROCESS / 服装贸易

Buying Plan

The first step in merchandising is the buying plan. Because the merchandiser is responsible for providing an inventory that reflects consumer demand, sizing, and season, while staying within a set budget, a buying plan is essential. Predicting needs, demand, and sizing is called fashion forecasting. The buying plan is normally done on a six-month basis and is called a six-month buying plan. The buying plan lists expected sales and last year's sales for each classification. A classification is a type of good, for example, knit tops or denim bottoms①. In addition, it also lists an open-to-buy figure. The open-to-buy is the dollar amount of merchandise the buyer may purchase during a given period②. For all practical purposes, the open-to-buy figure is the money a buyer has been allocated to spend on purchases. This figure is based on expected sales figures, desired inventory levels, and items already on order. Open-to-buy provides a guideline for buyers by making sure they do not over-purchase or under-purchase③.

Buy and Order

Once knowing what he or she has to spend, the buyer will then work with designers, depending on the type of company, or attend trade shows to place orders for new fashions. It is imperative that the buyer has an accurate plan so that he or she doesn't over-buy or under-buy during the trips. Most buyers will make at least two trips a year, one for fall/winter and one for spring/summer. There are two types of places buyers can go to purchase goods. They can attend trade shows, normally located in fashion centers. A fashion center is a city known as place in which many trade shows or showrooms are located. There are also specialty trade shows, such as for wedding apparel. A showroom is the second place a buyer can go to order merchandise. Showrooms are normally run either by designers or sales representatives of the clothing line.

Buyers may also have sales representatives from various manufacturers visit them with line boards and catalogs. The buyer, as expected, has performed extensive research on trends, knows the image the company is trying to portray, and knows the price range of goods sold within his or her store④. As a result, when buying and ordering, the buyer will keep these things in mind.

Receive Orders and Provide Information to Staff

Depending on the type of company, the buyer or the merchandise planner will be responsible for making sure the items arrive at the store. The items that end up in the stores should be appropriate to climate and size of the store. In addition, the buyer will communicate information about new items coming in to the visual merchandising team as well we to the people on the sales floor⑤. This can be performed informally or formally by new product meetings with salespeople.

Monitor Inventory Levels and Sales

Depending on the company, monitoring of inventory might be the job of the merchandise planner. Stores use sophisticated scanning devices and inventory software to take inventory in their stores, and usually it is ongoing⑥. This software can prevent theft, but it also provides real-time information to buyers about inventory levels. In any size company it is important for the buyer to know how well items are selling in the stores. For items that are selling very well, the buyer may want to reorder the merchandise⑦. For those that are not selling well, the buyer may want to transfer the merchandise to a different store or markdown the merchandise. Most retailers have sophisticated computer systems to monitor inventory levels.

Negotiate Buybacks or Advertising with Vendors

The final responsibility of the merchandiser is to negotiate with vendors. It is important to note, though, that negotiation will occur throughout the fashion buying process. The buyer will negotiate price during the buying and ordering process, but there are a couple of other things he or she will negotiate as well. First, for companies that advertise, the buyer may work with the marketing department to negotiate with the vendor to split the cost of the advertisement, if it is featuring the garment of the manufacturer⑧. Another negotiation might be to buyback goods that did not sell. An example might be a selection of lime-colored shirts that the vendor promised would be "hot" then didn't sell.

In addition, a merchandiser may negotiate use of point-of-purchase displays provided by the vendor at no cost or a reduced cost.

MERCHANDISER

In many companies the apparel merchandiser plays a key role in the sourcing function. In small companies the merchandiser may have total responsibility for sourcing decisions; midsized companies may have buyers, production specialists, or sourcing agents who report to the merchandiser; larger companies may have a vice president of product, manufacturing, or sourcing and staff offices in the countries manufacturing their products.

Whatever the company structure, it is imperative that all those involved in the merchandising function understand the mechanics of global sourcing. There is a complex interrelationship between style development, fabric sourcing, and product sourcing. As the variety of fabrics and trims and the international sources of fabrics increase, the merchandiser is frequently involved in the complicated sourcing decision because fabrics are sourced from one country, trims from another, and production from yet another. This creates timing and logistics challenges, which are part of the overall sourcing decision. Merchandisers often develop whole segments of a line based upon materials available from the sourcing network.

No matter what role the merchandiser and product manager play in sourcing, from direct line responsibility to coordination with a specialized sourcing or production executive, merchandising and sourcing share a common goal-get the right product to the customer at the right price and at the right time.

Words and Phrases

inventory ['invəntəri] *n.* 报表,清单
knit tops 针织上衣
denim bottoms 粗斜纹劳动布下装,牛仔裤类服装
open-to-buy (OTB)进货限额,许购定额
allocate to 分派,配给
guideline ['gaidlain] *n.* 指导方针
over-purchase 溢价购置
under-purchase 低价购置
imperative [im'perətiv] *adj.* 绝对必要的,不可避免的
over-buy 溢价购买
under-buy 低价购买
showroom ['ʃəurum] *n.* (样品)陈列室,展览室
specialty ['speʃəlti] *n.* 专长,专业,特点
representative [ˌrepri'zentətiv] *n.* 代理
visual merchandising 展示销售
salespeople ['seilzˌpi:pl] *n.* 售货员,店员
inventory software 资产管理软件
real-time 实时的,快速的
markdown ['ma:kdaun] *n.* 减价商品,削价商品
inventory level 存货水准,库存水平
negotiate [ni'gəuʃieit] *v.* 议定,商定,谈判
buyback [bai'bæk] *v.* 回购,产品返销
lime-colored 酸橙绿色的
point-of-purchase 卖点
display [dis'plei] *n.* 展览,陈列 *v.* 展示,陈列,展出
midsize ['midsaiz] *adj.* 中等大小的,中号的,中型的
specialist ['speʃəlist] *n.* 专家,行家
sourcing agent 采购代理
vice president 副总统(或大学副校长等)
imperative [im'perətiv] *adj.* 必要的,紧急的,极重要的,迫切的,急需处理的
interrelationship 相互关系,联系,影响;干扰
logistics [ləʊ'dʒistiks] *n.* 后勤;物流
segment ['segmənt] *n.* 部分,份,片,段
responsibility [riˌspɔnsə'biliti] *n.* 责任,义务,负担;可靠性
coordination [kəʊˌɔ:din'eiʃən] *n.* 协调,和谐

Notes

① A classification is a type of good, for example, knit tops or denim bottoms.
一个类别就是一种产品,例如针织上装或牛仔裤。

② In addition, it also lists an open-to-buy figure. The open-to-buy is the dollar amount of merchandise the buyer may purchase during a given period.
此外,它还包括许购定额。许购定额是指采购商在一定时间内可以用来购买产品的总额度。

③ This figure is based on expected sales figures, desired inventory levels, and items already on order. Open-to-buy provides a guideline for buyers by making sure they do not over-purchase or under-purchase.
许购定额是根据预期的销售额、理想的库存水平和已有的订单而制定的。许购定额为采购商提供了业务标准,避免购买过量和购买不足。

④ The buyer, as expected, has performed extensive research on trends, knows the

image the company is trying to portray, and knows the price range of goods sold within his or her store.
理想的采购商,需要对流行趋势有深刻的研究,了解企业发展的方向,了解商场产品价格的最佳范围。

⑤ In addition, the buyer will communicate information about new items coming in to the visual merchandising team as well as to the people on the sales floor.
此外,采购商还需与新产品和视觉陈列团队以及销售部门进行沟通。

⑥ Stores use sophisticated scanning devices and inventory software to take inventory in their stores, and usually it is ongoing.
商场常连续地使用复杂的扫描设备和资产管理软件。

⑦ For items that are selling very well, the buyer may want to reorder the merchandise. For those that are not selling well, the buyer may want to transfer the merchandise to a different store or markdown the merchandise.
对于那些销售良好的产品,采购商会考虑翻单,而那些滞销的产品,采购商则会计划分销或打折销售。

⑧ First, for companies that advertise, the buyer may work with the marketing department to negotiate with the vendor to split the cost of the advertisement, if it is featuring the garment of the manufacturer.
首先,采购商会协同销售部门就公司的广告与卖家进行协商,目的是削减广告成本,并保证广告能体现制造商服装的特点。

Discussion Questions

1. Why do product developers expect their sourcing partners to take ownership of the textiles specified for apparel products?
2. Research textile periodicals and the Internet for new fabric and fiber developments. How might these developments influence fashion?

EXTENSIVE READING

WHAT IS FASHION SALES PROMOTION?

Fashion sales promotion consists of all the functions and activities within a store that are developed and used to influence the sales of its fashion merchandise. These activities include personal selling, displays, advertising, fashion shows, and many other supplementary selling activities designed to increase sale of apparel.

Successful fashion sales promotions do many other things also. They build continuing customer loyalty to the particular store and they communicate the image that your store wishes to be known among its customers. Sales promotion efforts enlighten the public about new trends in fashion; inform the public of advanced methods of merchandising that will make it easier and more convenient to shop; and announce special merchandise events or special prices. The promotion is always aimed at attracting more customers and new groups of customers who may not have been shopping in the store. Fashion promotions frequently attempt to establish and continue to enforce a particular store's fashion leadership and authority. In effect, the store is saying: "See us first for all your apparel needs!"

图 42 While many of our styles today are similar, they come in a variety of colors, designs, and fabrics

Who is responsible for planning fashion sales promotion events depends on the size of the organization. Larger businesses usually create special departments to plan promotions and assign the responsibility to specially trained individuals. These people work with all the departments in the store-merchandising, advertising, display, and publicity departments. The jobs are variously known as public relations, advertising and promotion, and so forth, but all are creative positions with the responsibility of promoting the store and its fashion apparel. In smaller stores, the promotion activities may be planned by the store manager or assistant. Actually, any store associate may present a sales promotion plan for consideration.

Words and Phrases

promotion [prəˈməuʃən] *n.* 推广，推销，促销

supplementary [ˌsʌpliˈmentəri] *adj.* 补充的

loyalty [ˈlɔiəlti] *n.* 忠诚

enlighten [inˈlaitən] *v.* 启发，开导

inform [inˈfɔːm] *v.* 通知，告诉

convenient [kənˈviːnjənt] *adj.* 方便的，便利的

announce [əˈnauns] *v.* 通知，预告

enforce [inˈfɔːs] *v.* 加强，强化

LESSON 26 STANDARDS / 标准

Apparel companies that have recognized the need to instill quality throughout their business operations have established expectations for the garments they produce and sell①. Standards are descriptions of acceptable measures of comparison for quantitative or qualitative value that are communicated to all involved in the development and execution of a product②. These descriptions are expectations for a variety of apparel factors such as dimensions, materials, style components, appearance, and performance. These factors apply to raw materials, design, production and packaging. It is critical that standards be expressed in language that is clearly understood by all personnel that could affect the quality of a garment style. At times this requires the use of illustrations, photographs, diagrams, definitions, and samples.

Quality expectations should be communicated in written form such as a quality assurance manual that is distributed to company personnel and outside suppliers and contractors. Quality assurance manuals should state the precise and exact characteristics of a product including minimum levels of performance and expectations.

Standardization is the process of establishing rules for compliance with the standards a company has developed for its products. Standardization requires cooperation among all company personnel as well as suppliers, contractors, and consumers③. When an entire industry adopts a standard, it is referred to as an industry standard. When the need for a standard exists but not all companies choose to develop or follow that standard, it is known as a voluntary standard④. When the issue of public safety is raised, for example, in the case of the flammability of infant's sleepwear, the use of drawstrings, or the use of small parts, the government becomes the primary partner in the process. If the standard is one that is required by statute, it is known as a mandatory standard⑤.

Before any standards are acceptable to all the affected participants, they will be examined, evaluated, and probably revised several times. Since standards provide a means for the entire supply chain to evaluate and compare similar products, standards encourage competition. For example, in sewing the center placket of an expensive shirt 22 stitches per inch is used as compared to using 10 or 12 stitches per inch for sewing a less expensive shirt⑥. The merchandiser must pay careful attention to the role of standards and standardization in the development and manufacture of their product lines.

Words and Phrases

instill [inˈstil] *v*. 滴注，慢慢地灌输
description [disˈkripʃən] *n*. 说明书
comparison [kəmˈpærisən] *n*. 比较，对照
execution [ˌeksiˈkjuːʃən] *n*. 执行，实行
diagram [ˈdaiəgræm] *n*. 图表
definition [ˌdefiˈniʃən] *n*. 定义
precise [priˈsais] *adj*. 精确的，明确的
standardization [ˌstændədaiˈzeiʃən] *n*. 使合标准；使标准化
supplier [səˈplaiə] *n*. 厂商，供应商
contractor [kənˈtræktə] *n*. 承包商
industry standard 行业标准
voluntary standard 自愿执行标准
flammability [flæməˈbiləti] *n*. 易燃，可燃性
infant [ˈinfənt] *n*. 婴儿，幼童
drawstring [ˈdrɔːstriŋ] *n*. 拉带，细绳
mandatory standard 法定标准
statute [ˈstætjuːt] *n*. 法令，法规
participant [paːˈtisipənt] *n*. 参加者
placket [ˈplækit] *n*. 口袋

Notes

① Apparel companies that have recognized the need to instill quality throughout their business operations have established expectations for the garments they produce and sell.

服装公司已经意识到整个商业操作中质量管理的重要性，并设立相应的生产和销售的预期值。

② Standards are descriptions of acceptable measures of comparison for quantitative or qualitative value that are communicated to all involved in the development and execution of a product.

标准是对于在产品研发和制作过程中广泛交流的定量或定性的数值，建立的可接受的程度的描述。

③ Standardization requires cooperation among all company personnel as well as suppliers, contractors, and consumers.

标准化需要供应商、经营商和消费者达成一致。

④ When an entire industry adopts a standard, it is referred to as an industry standard. When the need for a standard exists but not all companies choose to develop or follow that standard, it is known as a voluntary standard.

当整个行业接受同一个标准的制约，这个标准被称为是行业标准，当某个标准被设定，但不是所有的公司都遵循这个标准，此标准被称为是自愿执行标准。

⑤ When the issue of public safety is raised, for example, in the case of the flammability of infant's sleepwear, the use of drawstrings, or the use of small parts, the government becomes the primary partner in the process. If the standard is one that is required by statute, it is known as a mandatory standard.

在引发公共安全问题时，例如考虑到婴儿睡衣的可燃性、抽带和小部件的使用等，政府成为标准执行中的首要角色。如果标准是法定的，则被称为是强制性标准。

⑥ Since standards provide a means for the entire supply chain to evaluate and compare similar products, standards encourage competition. For example, in sewing the center placket of an expensive shirt 22 stitches per inch is used as compared to using 10 or 12 stitches per inch for sewing a less expensive shirt.

由于标准为整个供应链提供一个评估和比较同类产品的方法，因此标准鼓励竞争。例如，前中口袋的缝制：高级衬衫采用每英寸 22 针的缝制工艺，而每英寸 10 到 12 针则用于缝制较廉价的衬衫。

Discussion Questions

1. How can a manufacturer predict how many garments to make?

EXTENSIVE READING

STANDARD ORGANIZATIONS

There are several national and international organizations that develop and promote standards used within the textile and apparel industries. The following four organizations are key contributors to establishing quality standards and testing procedures that affect the apparel industry:

The American Association of Textile Chemists and Colorists (AATCC) is recognized around the world for its standard methods of testing dyed and chemically treated fibers and fabrics, which measure such performance characteristics as colorfastness to light and laundering, durable press, dimensional stability, and water repellency. Today, practically all the dyes and finishes and many chemicals produced in the United States are evaluated by AATCC methods.

The American Society for Testing and Materials (ASTM) is the largest nongovernmental organization in the world that writes standards for materials used in many industries. ASTM Committee D-13 is responsible for establishing and maintaining textile and apparel standards and specifications. ASTM is the foremost developer of voluntary consensus standards, related technical information, and services having internationally recognized quality and applicability that:

- Promote public health and safety.
- Contribute to the reliability of materials, products, systems, and services.
- Facilitate national, regional, and international commerce (ASTM).

The American National Standards Institute (ANSI) maintains as its primary goal the enhancement of global competitiveness of U.S. business and the American quality of life by promoting and facilitating voluntary consensus standards and conformity assessment systems and promoting their integrity. Many ASTM procedures have been accredited through the ANSI certification program.

The International Organization for Standardization (ISO) is composed of national standards institutes from large and small countries, industrialized and developing countries, in all regions of the world. ISO develops voluntary standards that represent an international consensus in state-of-the-art technology.

Words and Phrases

contributor [kən'tribjuːtə] *n*. 贡献者

establish [is'tæbliʃ] *v*. 成立,建立,设立;创立;开设,确立

procedure [prə'siːdʒə] *n*. 过程,步骤程序

launder ['lɔːndə] *v*. 洗[烫]衣

dimensional stability 尺寸稳定性

water repellency 拒水性

nongovernmental ['nɔn'gʌvənməntəl] *adj*. 非政府的,非政治(上)的

foremost ['fɔːməust] *adj*. 最初的,最先的,第一流的

consensus [kən'sensəs] *n*. (意见等的)一致

internationally [ˌintə'næʃənəli] *adv*. 国际性地,在国际间

applicability [ˌæplikə'biləti] *n*. 适用性,适应性

contribute to 捐献,贡献

reliability [riˌlaiə'biliti] *n*. 可靠性,安全

性；可信赖性

enhancement [in'ha:nsmənt] *n*. 增强，促进，提高

competitiveness [kəm'petitivnis] *n*. 竞争

facilitate [fə'siliteit] *v*. 使……容易，使不费力

conformity [kən'fɔ:miti] *n*. 相似；遵从，顺从

assessment [ə'sesmənt] *n*. 评估

integrity [in'tegriti] *n*. 诚实，正直，廉正

accredited [ə'kreditid] *adj*. 可接受的；可信任的，被正式认可的

certification [ˌsə:tifi'keiʃən] *n*. 证明，保证，鉴定证明书

state-of-the-art 达到最新技术发展水平的

LESSON 27 CHECKING THE QUALITY OF BULK PRODUCTION / 产品检验

Most fashion retailers supply their own quality manuals to manufacturers to ensure that the standards they require for their products are crystal clear to both parties①. Retail garment technologists may be responsible for writing, reviewing or updating the company's quality manuals, taking into account changes in the law or the market at which the products are aimed. Quality manuals cover legislation, health and safety issues and environmental considerations. With the increase in overseas sourcing quality manuals may be adapted to reflect the working practices in different countries and to ensure clear communication. Indications of poor standards which garment technologists watch out for when inspecting garments include:

- Puckered or inconsistent stitching quality.
- Seam slippage.
- Security of fastenings and trims.
- Position of pockets or fastenings.
- Loose threads.

Garment technologists working for suppliers are responsible for checking the quality standards of garments during production and sending samples of these garments to the retailer if required②. Fabric testing is part of the quality control process for fashion retailers. Checking that the testing standards for fabrics have been met by the supplier is part of the retail garment technologist's role-or the fabric technologist's , if the company has one.

Retail garment technologists sometimes travel to visit suppliers to observe garments being manufactured. These visits are known as an "in-work checks" and involve assessing the quality standards of products which are currently being manufactured or are ready for delivery. This can help to save the time and expense of returning low-quality merchandise to the manufacturer if the garments are rejected③. Quality inspections include measuring various dimensions of garments to ensure that they meet the required specifications and assessing the quality of manufacture. As fabric is a flexible material a slight variation of 0.5cm to 1cm from the specified measurement is allowed depending upon the part of the garment, referred to as "tolerance". It is impossible for retail garment technologists to see all of the styles for which they are responsible in production, particularly as so many are manufactured overseas. This has led to an increase in self-certification by suppliers, working to a specified inspection system, giving them more responsibility for adhering to the retailer's quality standards in production④. Independent testing labs can also be contracted to carry out in-work checks on behalf of the retailer⑤.

Finished garments are transported to the retailer's distribution centre (DC), which is often based in one central location or several regional sites. Retail garment technologists often travel to the DC to check the quality of stock prior to its delivery to stores. Retailers often employ QCs based permanently at the DC who liaise with the garment technology department at head office⑥. QCs inspect a small percentage of garments at random. If a quality problem is anticipated, some styles may require 100 percent inspection at the warehouse and if the quality is consistently low, the delivery could be rejected.

Words and Phrases

update [ʌpˈdeit] *v*. 修正,更新
quality manual 质量手册
legislation [ˌledʒisˈleiʃən] *n*. 法规
environmental consideration 有关环境方面的考虑
watch out for 密切注意,戒备,提防
inspect [inˈspekt] *v*. 审查,检查
pucker [ˈpʌkə] *v*. 折叠,起皱
inconsistent [ˌinkənˈsistənt] *adj*. 断断续续的,不连贯的
seam slippage 跳线
slight [slait] *adj*. 细小的,轻微的
self-certification 自行认定
adhere to 坚持,遵守
independent testing labs 独立的测试实验室
distribution centre (DC)配送中心
quality control (QC)质量控制
liaise [liːˈeiz] *v*. 建立联络关系、取得联系
at random 随机地
warehouse [ˈwɛəhaus] *n*. 仓库
delivery [diˈlivəri] *n*. 交货,出货

Notes

① Most fashion retailers supply their own quality manuals to manufacturers to ensure that the standards they require for their products are crystal clear to both parties.
许多时装零售商为生产厂家提供质量手册,确保双方都非常清楚产品的指定标准。

② Garment technologists working for suppliers are responsible for checking the quality standards of garments during production and sending samples of these garments to the retailer if required.

供应商的服装工艺师在制作中负责检查成衣的质量标准，并按照零售商的要求寄出样衣。

③ These visits are known as an "in-work checks " and involve assessing the quality standards of products which are currently being manufactured or are ready for delivery. This can help to save the time and expense of returning low-quality merchandise to the manufacturer if the garments are rejected.
这个验货被称为跟单，包括对正在生产和准备销售的产品质量标准的评估。跟单有助于降低被拒次品的回单率，节约时间和成本。

④ This has led to an increase in self-certification by suppliers, working to a specified inspection system, giving them more responsibility for adhering to the retailer's quality standards in production.
这导致了供应商自行认证的增加，采用一个专门的检查系统，使供应商有更多的责任感，使产品能够符合零售商的质量标准。

⑤ Independent testing labs can also be contracted to carry out in-work checks on behalf of the retailer.
独立的测试实验室也能签订合同，代表零售商去执行跟单检查。

⑥ Retailers often employ QCs based permanently at the DC who liaise with the garment technology department at head office.
零售商常常聘用销售中心资深的质量控制员，他们与总部技术部门有一定的联系。

Discussion Questions

1. What colors will sell best in different parts of the country?

EXTENSIVE READING

QUALITY CONTROL

Making sure that every garment produced is of the appropriate and agreed-upon size specs and quality level is a bigger challenge than you might realize! There are numerous components and operations required for even the simplest garments. With so much pressure on keeping prices low, everyone in the process is trying to use the least expensive but still acceptable materials, put together in the shortest possible time. It is a very fine line between acceptable and unacceptable quality. Of course, in high-quality lines, only the best materials are used. But even there, quality control personnel need to stay alert.

Most manufacturers and wholesalers entrust this responsibility only to very experienced, technically trained pros, don't expect a startup job here! Chances are that a quality control (QC) professional will have an engineering degree plus significant production-line experience. They would be entrusted to:

Evaluate yarn and fabric mills for suitability.

Evaluate garment production plants to see if the required quality level can be achieved. They can also tell how efficiently the plant is working.

Know the capabilities of spinning, knitting, dying, cutting, and sewing machinery.

After all this preliminary investigatory work, QC will typically inspect piece goods before they are cut and monitor the garment production to be sure that all measurements and other requirements are being met. Most importers require that their own QC person or agent issue a certificate of inspection before the goods can be shipped.

Tests for shrinkage, seam strength, colorfastness, and other points of performance will also be conducted by QC.

Words and Phrases

greed-upon 互相认可的

numerous [ˈnjuːmərəs] *adj*. 许多的，无数的

high-quality 高级的

personnel [ˌpəːsəˈnel] *n*. 人员，职员

alert [əˈləːt] *adj*. 警惕的；警觉的，留心的

wholesaler [ˈhəulseilə] *n*. 批发商

entrust [inˈtrʌst] *v*. 委托，托付

startup [ˈstaːtˌʌp] *n*. 启动

significant [sigˈnifikənt] *adj*. 有意义的，重要的，重大的

production-line 生产线，流水线

evaluate [iˈvæljueit] *v*. 评估，评价，赋值

suitability [ˌsjuːtəˈbiliti] *n*. 合适，适当，相配，适宜性

efficiently [iˈfiʃəntli] *adj*. 能胜任的，有能力的，效率高的

investigatory [inˈvestigeitəri] *adj*. 研究的，好研究的

inspect [inˈspekt] *v*. 检阅，检查，审查，视察

piece goods 匹头，布匹

agent [ˈeidʒənt] *n*. 代理人，代理商

certificate [səˈtifikit] *n*. 证(明)书，凭证

performance [pəˈfɔːməns] *n*. 履行，实行，执行，完成

conduct [ˈkɔndʌkt] *n*. 管理，处理

LETTERS OF CREDIT

Letters of credit (L/Cs) are financial agreements between the sourcer (buyer) and its bank (issuing bank) to transfer responsibility for paying the seller (foreign contractor) to the issuing bank. This form of guaranteed payment based upon the credit worthiness of the issuing bank is the primary financial instrument used in international trade. Bank charges may include 0.25 percent to 1 percent of the total transaction price for issuing fees and negotiation fees, and flat fees per occurrence for discrepancies and amendments. A discrepancy requires that the bank investigate any difference between the documentation required by the L/C and the documentation presented. An amendment is any change made to the L/C after it is issued.

Application and Agreement for Irrevocable Commercial Letter of Credit
To：Wachovia Bank，National Association（"Bank"）

Please TYPE information in the fields below. We reserve the right to return illegible applications for clarification.

Date：	12/1/09	Please issue an irrevocable commercial letter of cresit substabtially as set forth below and forward same through a selected correspondent by： ■ Teletransmission □Overmight Carrier □Mail □Other If Other，explain ______________
L/C#：	(Bank Use Only)	Advising Bank Name (optional)：______________

Applicant(Full Name & Address) UPspace Sportswear Company School House Lane and Herny Avenue Philadephia PA 19144	Currency and Amount in Figures： $ 1,000,000 Tolerance Amount (if applicable)：1% Currency and Amount in Words： one million U.S. Dollars
Beneficiary(Full Name & Address) High Quality Garment Company Ltd. Apparel Drive Bangkok Thailand	Expiration Date： 6/15/10 Latest Shipment Date： 5/1/10

Draft Tenor：■Sight；OR ________ Days from □Sight or form □Bill of Lading Date：OR □Other ________
Draft For：■100% OR □ ______% of the invoice value，drawn at the Bank's option，on the Bank or its correspondent.
Charges：Wachovia's charges are for the： □Applicant ■Beneficiary
Discount：Discount charges，if any，are to be paid by the： ■Applicant □Beneficiary

Shipment From(Port of Loading)： Bangkok，Thailand	Shipment To(Port of Discharge)： Baltimore，MD

Brief Merchandise Description：ten thousand pair of women's cotton athletic pants

Terms of Shipment： ■ FOB □C&F(CFR) □CIF □FAS □Other ________
Letter of Credit to be：■ Transferable □Non-Transferable
Partial Shipments are：■Permitted □Non-Permitted
Transshipments are：□Permitted ■Non-Permitted

DOCUMENTS REQUIRED			
Transport Document(select one)	Freight	Notify Party：	
■Full Set Clean Multi Modal	□Collect	Company Name：	UPspace Sportswear Company
Transport bill of lading	■Prepaid	Contact name：	David Jean
□Full Set Clean on Board Marine	□Issued or endorsed to the order of the issuing Bank.	Address：	School House Lane and Herny Avenue Philadephia PA 19144
■Air Waybill		Phone	(215)951-00000
□Other ______		Fax：	(215)951-11111

Insurance(select one)：

■Insurance effected by Applicant. No insurance document is required.

□Air or Marine/War Insurance Policy or Certificate Covering "All Risks" for 110% Invoice cost. Specify other risks as needed：

Chapter 7 COMPUTER AIDED DESIGN SYSTEM / 计算机辅助服装设计

LESSON 28 COMPUTER AIDED DESIGN SYSTEM / 纸样绘制系统

Computer-Aided Design is one of the many tools used by engineers and designers and is used in many ways depending on the profession of the user and the type of software in question. There are several different types of CAD. Each of these different types of CAD systems requires the operator to think differently about how he or she will use them and he or she must design their virtual components in a different manner for each.

Pattern Generation Software

Pattern generation software is pattern design software that operates in a question-and-answer format. Rather than drafting a new collar on a shirt using drawing tools, the patternmaker uses word and measurement commands to modify the style on the screen①. Many different garment components are stored in the system and the designer can combine and alter the parts to conform to the sketch. For example, Bodice Front and Back X can be called up and combined with Sleeve Y and Collar Z. If the shoulder seam needs reduction, the operator can tell the computer to reduce by 0.5 inches and the computer will not only reduce the bodice pieces, but it will also make the appropriate adjustments to the sleeve and collar. Here again, fit has already been tested on all of the stored pattern parts, so sample-making procedures can be shortened or eliminated. After the pattern has been completed for the new design, the computer can generate a complete set of patterns in all sizes.

Grading Systems

After a sample size pattern has been input, it has to be graded up and down in size. Certain points on the pattern are considered "growth points" or places at which the pattern has to be increased or decreased to accommodate changing body size. At each growth point, the operator indicates a grade rule to the computer. This grade rule lets the system know which way to move on the X and Y coordinates in order to increase or decrease size. The system will then automatically produce the pattern shapes in all the prespecified sizes. The pattern parts can then be placed one inside another, or nested, from smallest to largest so that the patternmaker can pick up any problems with the grade at a glance.

Marker-Making Systems

A marker is the arrangement of all the pattern parts for a particular style in the optimal configuration for maximum fabric utilization. In marker-making systems, the patterns in all graded sizes appear on the screen in small scale and the operator moves them around until the least percentage of fabric waste is achieved[2]. Each pattern size appears in a different color to avoid confusion. The system will let the operator know the percentage of waste by command. Some systems use algorithms to compute pattern arrangement and the operator's involvement is minimized. Width parameters of the fabric are taken into account as there are any stripe or plaid lines and nap considerations[3]. Automatic matching of stripes and plaids is common to most systems. If the operator wants to rotate the patterns by degree to allow them to be squeezed on more tightly, this can be accomplished easily.

Plotting functions are linked to marker-making systems and allow the marker to be printed in varying scales on large rolls of continuous paper. The paper is then overlaid on the stacked fabric prior to cutting. Individual patterns in full scale or half scale as well as nested patterns can be printed out using the plotter[4].

Cutting Operations

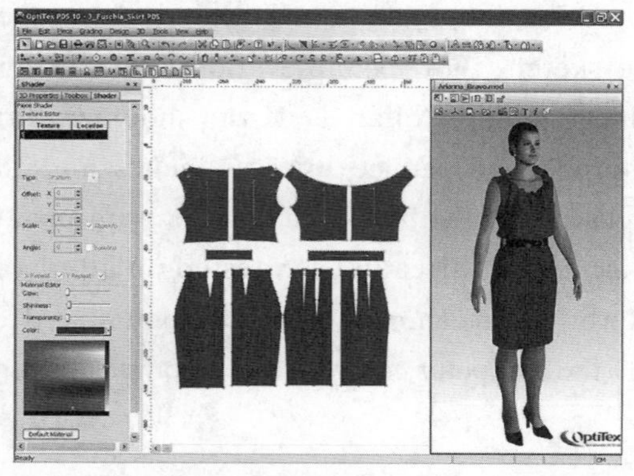

图 43 Image rendered with computer aided design

Cutting operations are connected to marker-making systems and calculate the number of layers of fabric that have to be cut to accommodate orders for color and size in a particular style. Cutting systems can be used without plotting out the marker, but rather by passing along the information about pattern shapes generated by the marker-making system directly to an automated cutting machine. The cutting machine operates on its own without being pushed by a human being. Cut-paths for the machine to follow can also be calculated by the computer for the most efficient and accurate results.

Words and Phrases

optimal [ˈɔptiməl] *adj*. 最佳的，最理想的

configuration [kənˌfigjuˈreiʃən] *n*. 结构；构造；布置方位；构型，排列

utilization [ˌjuːtilaiˈzeiʃən] *n*. 利用，使用，应用

marker-making system （服装）排料系统

confusion [kənˈfjuːʒən] *n*. 混乱，混淆

algorithm [ˈælgəriðəm] *n*. 算法；规则系统；编码

arrangement [əˈreindʒmənt] *n*. 整顿，布

置，排列

involvement [in'vɔlvmənt] *n.* 卷入，介入

consideration [kənˌsidə'reiʃən] *n.* 考虑，思考

automatic [ˌɔːtə'mætik] *adj.* 自动的

rotate [rəu'teit] *v.* 旋转；循环，自转轮换

squeeze [skwiːz] *v.* 挤，塞，压进，挤入

accomplish [ə'kɔmpliʃ] *v.* 完成，贯彻，实行

plot [plɔt] *v.* 绘制平面图

roll [rəul] *v.* 卷成，卷拢，卷起

prior to 在前，居先，在……之前

nested pieces 嵌套式样片

plotter ['plɔtə] *n.* 描绘器，图形显示器，绘图器；标图员

cut-path 裁剪轨迹

indicate ['indikeit] *v.* 指示；指出，表明；显示；象征

coordinate [kəu'ɔːdineit] *v.* 坐标

prespecified [pri'spesifaid] *adj.* 预定的，预先设计好的

sample-making 样品制作

pattern generation software 纸样生成（绘制）软件

question-and-answer 问答式

combine [kəm'bain] *v.* 使联合，合作

alter ['ɔːltə] *v.* 修改，改动；改变

conform to 符合，遵照

adjustment [ə'dʒʌstmənt] *n.* 调整，调正，整理；修正

procedure [prə'siːdʒə] *n.* 过程，步骤程序

eliminate [i'limineit] *v.* 除去，排除，删除

Notes

① Pattern generation software is pattern design software that operates in a question-and-answer format. Rather than drafting a new collar on a shirt using drawing tools, the patternmaker uses word and measurement commands to modify the style on the screen.
纸样生成软件是一个问答模式的纸样设计软件，与使用绘图工具绘制一个新的衬衫领不同，纸样设计师通过输入文字和尺寸要求在屏幕上设计款式的纸样。

② A marker is the arrangement of all the pattern parts for a particular style in the optimal configuration for maximum fabric utilization. In marker-making systems, the patterns in all graded sizes appear on the screen in small scale and the operator moves them around until the least percentage of fabric waste is achieved.
排料就是针对某款服装的所有裁片以获得最小损耗率为目的而进行的最优化方式排列，在排料系统中，操作者在屏幕上移动所有尺码的裁片小样，以获得最小的损耗率。

③ Width parameters of the fabric are taken into account as there are any stripe or plaid lines and nap considerations.
条纹布或格子布以及倒顺毛面料在排料时需要考虑面料的幅宽。

④ Plotting functions are linked to marker-making systems and allow the marker to be printed in varying scales on large rolls of continuous paper. The paper is then overlaid on the stacked fabric prior to cutting. Individual patterns in full scale or half scale as well as nested patterns can be printed out using the plotter.
绘图功能与排料系统相联，允许排料图在大型的滚筒纸上以不同的尺寸打印。这张

纸将被放在铺好的面料上进行裁剪。个别的全尺寸或者半尺寸的裁片和套裁裁片一样使用打印机打印出来。

Discussion Questions

1. How do you think the body scanner can be used in the apparel industry?

EXTENSIVE READING

BODY MEASUREMENT AND BODY SCANNING

SOFTWARE

Body measurement software uses specific individual measurements to modify a stored pattern and produce a new one that is particular to a single customer. Sometimes the customer's measurements at certain body points are fed into the system which then makes adjustments at matching points on the pattern. In other cases, a video image of the customer wearing a body stocking is used and the computer generates an image of the person and uses pattern generation software to convert measurements from the image to a specific pattern. These systems are used for tailored garments, perfect fit jeans, and by business that offer custom fitting. Another type of system that can be used for patternmaking and fitting allows for input of the dress form or mannequin. The computer can then lay the pattern parts over the form to see a simulation of the final garment. Matching of plaids and stripes and placement of pattern can be manipulated and visualized.

A new trend in customizing clothing includes technology called electronic body scanning. The technology that will allow fast and accurate body measurements to be taken is body scanning. A laser, white light, or other light source is used to illuminate the body, and cameras or sensors capture either specific body measurements or a three-dimensional digital image of the body. One of the drawbacks of current technologies is that individuals being scanned must either strip to their underwear or wear a full body stocking and strike specific poses during the scanning process.

Another technological hurdle for body scanning was transforming the digital data from the scanning process into customized two-dimensional patterns that will create a garment that fits perfectly. Fit criteria differ based upon varying garment styles and the mechanical properties of fabrics. A tremendous amount of research has been done in developing algorithms for different categories of garments and fabrics that can transform body dimensions into custom garment patterns. The development of appropriate software systems that can accommodate the many variables in creating accurate custom patterns will determine the future success of customization. The Textile and Clothing Technology Corporation continues as the price leader in developing technologically advanced body scanning. Their 3-D Body Scanner uses white light to scan the entire body in less than six seconds, and within a few minutes produces an exactly scaled 3-D model that accommodates the many variables in creating the accurate custom patterns that are required for mass customization.

Words and Phrases

body measurement software 人体测量软件

body stocking 女子贴身连衣裤，紧身连裤袜

mannequin ['mænikin] *n.* 服装模特儿；(橱窗里的)服装模型；人体模型

customize ['kʌstəmaiz] *v.* 定制，按规格改制，定做

body scanning software 人体扫描软件

smart card 智能卡

portable ['pɔːtəbl] *adj.* 便于携带的；手提式的，轻便的，可移动的

attune [ə'tjuːn] *v.* 调(音)，调谐使调和，使协调

customization [kʌstəmaɪ'zeʃən] *n.* 用户化，专用化，定制

personalization [ˌpəːsənəlai'zeʃən] *n.* 个性化

laser ['leizə] *n.* 激光

illuminate [i'ljuːmineit] *v.* 照亮

camera ['kæmərə] *n.* 照相机

sensor ['sensə] *n.* 传感器

capture ['kæptʃə] *v.* 捕获，捕捉

tremendous [tri'mendəs] *adj.* 巨大的

LESSON 29 SPECIFICATION AND COSTING SYSTEMS / 工艺单和成本软件系统

These systems store all style information including a flat sketch, size specifications, trim requirements, and size grade charts. These documents are called specifications sheets or spec sheets①. They keep track of all information pertinent to the design and production of a style including folding and shipping instructions if necessary. Spec and costing systems help to ensure accuracy and consistency when a company has a number of production sites, and are particularly helpful and even mandatory when a portion of the work is done overseas. Most companies have experienced communication problems related to cultural and language barriers and CAD visualizations minimize the risk of misunderstanding.

Specifications management is an extremely important aspect of CAD because it controls all of the paperwork that supports the production, costing, and delivery of the garment②. As a designer works on a garment, all of the costs related to that item, even the amount of thread consumption, can be determined immediately. The impact of any slight change in design or in a sewing sequence can be seen immediately. An apparel manufacturer's spec system is usually available on line so that access to information is available to anyone in the company, no matter where they are in the world.

Specifications and costing systems can be linked to inventory control systems as well. They add information from marker-making and preproduction systems and keep track of all data from yard goods and trims to the finished product. Specifications and costing systems connecting the mill, the factory, and the retail store expand the information network[③].

Commercial Software Systems

Software such as Adobe Photoshop, Adobe Illustrator, and many other commercially available graphics and paint programs are used by many companies to achieve specific goals, such as catalogue development, and as a supplement to proprietary systems they may own already. Some companies have come up with libraries either on disk or on CD ROM that an apparel manufacturer might purchase. These libraries contain hundreds of flat drawings or illustrations, historic costume examples, textile prints, and weaves or knit stitches.

Commercial pattern-making software is available for PCs and Macs that allows a small manufacturer or home or school user to create and modify patterns.

Words and Phrases

size grade chart　尺码表
keep track of　与……保持联系
pertinent [ˈpəːtinənt] *adj*. 有关的，相关的
consistency [kənˈsistənsi] *n*. 始终一贯；前后一致
barrier [ˈbæriə] *n*. 障碍
link to　把……和……联系起来
inventory [ˈinvəntəri] *n*. 财产等的清单；商品的目录；任何详细记载
preproduction [priːprəˈdʌkʃən] *adj*. 试生产
trim to　修饰使适合……
expand [iksˈpænd] *v*. 扩大，扩展，扩张
commercially [kəˈməːʃəli] *adv*. 商业上；通商上
available [əˈveiləbl] *adj*. 可用的，便于利用的，在手边的；适用于……的
graphics [ˈgræfiks] *n*. 图样；图案；绘图；图像
supplement [ˈsʌpliment] *n*. 增补(物)；补充(物)；添加物
proprietary system　专有系统
PC　所有装配 Windows 系统的电脑
Mac Macintosh　苹果电脑

Notes

① These systems store all style information including a flat sketch, size specifications, trim requirements, and size grade charts. These documents are called specifications sheets or spec sheets.
这个系统涵盖了款式的所有信息，包括平面效果图、尺寸规格、裁剪要求和尺码表，这些文件统称为规格表或工艺单。

② Specifications management is an extremely important aspect of CAD because it controls all of the paperwork that supports the production, costing, and delivery of the garment.
规格表管理是计算机辅助设计系统极为重要的一块内容，因为它涵盖所有生产、成本和出货的文件。

③ Specifications and costing systems can be linked to inventory control systems as well. They add information from marker-making and preproduction systems and keep track of all data from yard goods and trims to the finished product. Specifications and costing systems connecting the mill, the factory, and the retail store expand the information network.

规格表和成本核算系统也可以链接到库存控制系统。他们从排料系统和试生产系统中添加信息,并跟踪从购买面辅料开始直到最终产品的所有生产数据。规格表和成本核算系统将大大小小的工厂和零售商联接起来,扩展了信息网络。

Discussion Questions

1. Discuss the communication problems related to culture and language barriers between different fashion company.
2. What are the contents of a specification?

EXTENSIVE READING

QUICK RESPONSE

Success in the fashion and textile industries requires that the fabric producers supply the fashion industry, the fashion industry supplies the fashion retailers and the fashion retailers supply the consumer with what they want, when they want it. The demands of the consumer, the ultimate beneficiary at the end of this chain of events, along with the requirements of competition, have instigated a "speeding up" of the production and supply cycles in order to deliver goods even faster. This acceleration of response times to satisfy consumer demand involves the fiber and fabric industries, apparel manufacturers and fashion retailers all working as a team to efficiently speed up the delivery of fashion goods to the consumer. This innovation in the fashion and textile industries has been labelled "Quick Response" or QR. The QR concept was adopted in America from the 1970s onwards in an attempt to increase competition with imports. Since its introduction, QR has been accompanied by JIT (Just-in-Time), which is more applicable to the textile side of the process rather than the fashion or retail side. JIT is a form of inventory management; in effect, it prevents the unnecessary storing of materials used in manufacturing products. The upshot is that suppliers only deliver the required materials at the moment the manufacturer needs to make the product. Therefore, the on-costs involved with holding stock are kept to a minimum.

图 44 Garment product-o-rial system

Quick response is a true reflection of the way in which the fashion and textile consumer has transformed the structure of the fashion and textile industries from a "push system" to a "pull system".

Whereas originally fashion erred on the side of supply, pushing products on to the consumer, now it has shifted the emphasis to customer demand, using precise data to supply what the consumer needs. The role of technology, in supplying exact statistics to the appropriate links in the chain of supply to the consumer, is an integral part of the Quick Response methodology. At its conception in 1985, QR was the outcome of an amalgamation of a number of major American retailers, their suppliers and IBM. The Quick Response process was actualized using computer-to-computer transmission of sales data, which systematized the exchange of information between groups. Manufacturers, retailers and distributors were all kept informed of what was selling, where it was selling and what needed to be replenished. The results of this modification within the textile and fashion supply industries had a phenomenal effect on trade.

Words and Phrases

demand [diˈmaːnd] *n*. 需求,需要

beneficiary [ˌbenəˈfiʃəi] *n*. 受益人;受惠人;收款人

instigate [ˈinstiˌgeit] *v*. 煽动;唆使;鼓动

onwards [ˈɔnwədz] *adv*. 向前;前往

acceleration [ækˌseləˈreiʃən] *n*. 加速;加快

response [riˈspɔns] *v*. 回答;回音;答复,反应,响应

efficiently [iˈfiʃəntli] *adv*. 效率高地;有效地

speed up 加速

delivery [diˈlivəri] *n*. 投递,送交,传递,递送,交付

quick response 快速响应

competition [ˌkɔmpiˈtiʃən] *n*. 比赛,竞争;角逐

import [ˈimpɔːt] *n*. 进口,输入;输入的产品

accompany [əˈkʌmpəni] *v*. 陪伴,陪同,伴随……同时发生

just-in-time 及时

inventory [ˈinvəntri] *n*. 详细目录,存货清单,存货,库存;细目表

upshot [ˈʌpʃɔt] *n*. 结果

on-cost 间接成本;间接费用;附加行政费用

statistics [stəˈtistiks] *n*. 统计数据;统计资料

integral [ˈintigrəl] *adj*. 不可或缺的;作为组成部份的;完整的,完备的

methodology [ˌmeθəˈdɔlədʒiː] *n*. 原则方法学,方法论;方法论者,方法学者

conception [kənˈsepʃən] *n*. 思想,观念,概念;构想,设想

amalgamation [əˌmælgəˈmeiʃən] *n*. 合并;混合

actualize [ˈæktjuəlaiz] *v*. 实现,实施

transmission [trænzˈmiʃən] *n*. 传送,传播,传达,播送;传动装置,变速器

systematize [ˈsistimətaiz] *v*. 使系统化;使成体系

replenish [riˈpleniʃ] *v*. 补充,重新装满

phenomenal [fiˈnɔmənəl] *adj*. 非凡的,了不起的;惊人的;显著的

LESSON 30 FUTURE DIRECTION / 未来展望

Digital systems are poised to play a major role in the future of product development. As computer speed continues to increase, three-dimensional design systems and virtual reality video simulations will transform much, if not all, of the product development process into high-speed digital applications.

Realistic prototypes will be created using computer simulations, and advanced internet-based audiovisual communications will allow the product development team to evaluate them from anywhere around the globe①.

More and more CAD/CAM applications are including two-and three-dimensional drape simulation tools so that designers can create computer-generated images to enable them to evaluate styling concepts without having to manufacture prototypes. Textile CAD programs allow weave and knit generation of the fabrics used in creating even more realistic prototype simulations. The focus of CAD/CAM systems is on total production development solutions that provide suites of programs to accommodate all segments of the development process. These will include second generations of:

● Two-and three-dimensional creative drawing, modeling, and draping software to simulate style ideas as virtual prototypes for evaluation.

● Pattern design tools that can create two-dimensional garment patterns from three-dimensional simulations complete with accommodation for the mechanical properties of fabrics and ease allowances.

● Textile CAD systems that can create weave, knit, and print designs for application on virtual prototypes and then be electronically shared with weaving, knitting, and printing suppliers to shorten the fabric procurement cycle.

● Textile fabric digital printers that will allow actual prototypes to be created in fabric prints by the apparel designer.

● E-commerce B2B communications access to suppliers that will allow merchandisers and designers to view fabrics and findings inventories on-line, integrate those raw materials into virtual prototypes, and place raw materials orders electronically.

● Data management systems that will capture all specification details, drawings, and sketches and maintain data integrity for any new styles in progress while making the information available via the internet to the entire product development.

Collaboration among the members of the product development team is becoming more of a digital experience than a person-to-person one as the availability of low-cost videoconferencing over the internet is fast becoming a reality. A merchandiser in Los Angeles can link with designers in New York to view and discuss a garment prototype on a fit model in a sourcing partner's factory in Bangkok, Thailand②.

Through high-resolution video, data compression, and high-speed Internet transmission, the real time teleconference could allow the product development team to zoom in on the smallest construction detail or the smoothness of the garment drape on the model's body. Language translation software will be able to provide simultaneous translation in all languages. This will eliminate the language barrier that generates a large percentage of the problems associated with global apparel sourcing and deters collaborative product development.

图 45 AR advertising on electronics

Words and Phrase

internet-based 基于网络的

audiovisual [ˌɔːdiːəuˈviʒuːəl] *adj*. 视听的

computer-generated 计算机生成的

suite [swiːt] *n*. (软件的)套件

electronically [ilekˈtrɔnikli] *adv*. 电子地，电子操纵地

procurement [prəˈkjuəmənt] *n*. 获得，取得

e-commerce 电子商务

B2B Business to Business 企业到企业的电子商务模式

virtual [ˈvəːtjuəl] *adj*. 实际上的，实质上的

collaboration [kəˌlæbəˈreiʃən] *n*. 合作

person-to-person 个人之间的，面对面的

videoconferencing [ˌvidiəuˈkɔnfərənsiŋ] *n*. 视频会议

high-resolution 高分辨率，高清晰度

real time 实时的

teleconference [ˈtelikɔnfərəns] *n*. (通过电话，电视的)电信会议

simultaneous [ˌsiməlˈteinjəs] *adj*. 同时发生的，同步的

Notes

① Realistic prototypes will be created using computer simulations, and advanced

internet-based audiovisual communications will allow the product development team to evaluate them from anywhere around the globe.
使用计算机虚拟技术制作模拟样品，基于互联网技术的先进视听通讯系统，将支持来自世界各地的产品开发团队对样品进行评估。

② Collaboration among the members of the product development team is becoming more of a digital experience than a person-to-person one as the availability of low-cost videoconferencing over the internet is fast becoming a reality. A merchandiser in Los Angeles can link with designers in New York to view and discuss a garment prototype on a fit model in a sourcing partner's factory in Bangkok, Thailand.
当基于互联网技术的低成本视频会议快速变成一种现实时，产品开发团队之间的合作更像是数据的交流，而不是人与人之间的互动。洛杉矶的贸易商可以联系纽约的设计师，一起浏览和讨论泰国曼谷某合作工厂内试衣模特身上的一件样衣。

Discussion Questions

1. What is the future of the fashion business?
2. Which functions of the CAD cannot afford the needs of the fashion business?

EXTENSIVE READING

THE FUTURE OF FASHION INDUSTRY

Sustainability and digitization are the two major categories that will determine and change fashion in the coming years. No brand or retailer will be able to avoid them. Sustainability and digitization are anything but opposites. In fact, many sustainable developments can only be implemented through digital processes. These are the seven most important developments for the fashion of the future:

Future fashion: Focus on Health and Well-being

For years, textile manufacturers and brands have been experimenting with health-promoting properties of clothing, for example through integrated nano capsules that care for the skin and improve regeneration, or through intelligent threads that act as sensors, data conductors and power suppliers to generate heat over a wide area or measure moisture, pressure points and temperature.

In the home office, we have also learned that it is much more pleasant to be able to dress comfortably. Basically, fashion has been moving in an increasingly comfortable direction for decades. Sports and outdoor fashion are the main source of inspiration for this. The high acceptance of sneakers, stretch denim and men's jersey suits are all signs of this. People will not let these achievements be taken away from them again.

Mental health is also a typical health theme, which has received enormous tailwind due to the pandemic and is likely to boost the outdoor market in the long term. Being outdoors and experiencing nature is seen as a healing balance to the constant stress of our everyday lives.

Rental service and second-hand fashion for a lower carbon footprint

Instead of manufacturing or buying new products, more and more retailers, brands and consumers are turning to alternative consumption models. The second-hand market has been growing for years and is forecast to double to a total of 34 billion euros by 2025. More and more brands and retailers are investing in new resale business models and offering second-hand goods alongside their regular collections.

The same goes for rental services. In recent months, numerous retailers and brands have launched the new service. Whether it's luxury fashion, skiwear or outdoor equipment, the possibilities are many and Generation Z at the latest will be more interested in using products rather than owning them.

Reducing resource consumption with a circular economy

There are already a number of brands that are seriously addressing the issue of the circular economy, for example "On" with its subscription sneaker Cyclone or the cooperation between Bergan's of Norway and Spin nova. Most of these are lighthouse projects, and the number of functioning product cycles or even just recyclable products is still vanishingly small.

This is mainly because the entire value chain up to the disposal plant must pull together here. It is not enough to use recycled materials in production; the finished product must also be recyclable at the end of its life. Either because mono-material was used or because the individual parts can be easily separated from each other. However, the more robust and durable the products are, the more difficult it is to achieve the latter. Biological cycles — i.e., composting is also an option and are already being tested.

The outdoor sector, with its preference for synthetic material compositions and coatings, still has many challenges to solve. The industry is gearing up to develop new processes for recycling - also because this will soon be required by law.

Future fashion: Reduction of the CO_2 footprint

Organic is today, the future is regenerative. What this means, particularly in view of the climate targets, is that companies should not only strive to use sustainably produced raw materials, but increasingly also those that have been produced in a climate-positive way. They improve their own climate footprint. Since for apparel manufacturers their own products usually represent the largest item in their carbon footprint, it only makes sense to start here.

Using cotton as an example, this means only sourcing it from sources that use regenerative eco-agriculture methods that help pull carbon back into the soil.

Personalization: Specialists are in demand

Generalists are increasingly facing a credibility problem, because the days of "one fits all" are coming to an end. This by no means only means that the fit of clothing will be more closely oriented to "real" bodies in the future - for example, through body scanning technologies or personal avatars. The collections themselves will also be increasingly developed for and with precisely defined target groups and marketed in an equally targeted manner. In this context, the evolution of fashion ranges from gender-free collections to clothing brands that block cell phone radiation or offer menstrual underwear.

Digitized fashion: from digital influencers to the metaverse

With 3D technology, fashion is climbing to a new evolutionary level. More and more fashion companies are converting their product development to 3D. The technology has now reached the point where you can no longer tell the difference between a digitally generated product image and a photograph. This opens completely new possibilities: Digital products can be marketed online and tried on via personalized avatars before they have even been produced. Fashion shows and fashion campaigns can also be realized with digital models.

Finally, the virtual world of the Metaverse opens a completely new playground for digital fashion. Fashion brands can sell their products not only in the real world, but via NFTs in games like Roblox and Fortnite, because fashion plays an increasingly important role there, too. There have long been companies that only make digital fashion, and more and more classic fashion brands that are moving into the virtual world. Whether it's Nike with Nike land or Balenciaga, Ralph Lauren, Off-White and Karl Lagerfeld, the year 2021 has produced more new digital fashion worlds than ever before. And that is just the beginning.

Live shopping: digital shopping should become an event

Online shopping is growing, but it's honestly far from an experience. The solution: live shopping. In China, the trend is already a high-revenue mass phenomenon. There, 20 percent of online trade already takes place via live shopping, and the trend is rising. All this has only become established within the last five to six years.

Live shopping is seen by many as the future of online retailing because it has the character of an event and enables a direct, human encounter between retailer and consumer.

Conclusion: The fashion of future

The market provides room for innovation, and this paves the way for new ideas.

The clothing of the future is becoming more and more of a current trend and that makes people curious about new things.

Words and Phrase

nano capsule 纳米胶囊

regenerative eco-agriculture 可再生生态农业

personal avatar 个人头像

Metaverse *n*. 元宇宙

NFT(Non-Fungible Token) 非同质化代币

参考译文

第一课 服装

服装,是指人体上的包裹物或者成衣。布料和服装这两个词的意思是相关联的,前者是指面料或纺织品,后者是指人体上的成衣。最早的衣物不是用织物来制成的,而是用皮革或其他非织物制作的,这些非织造衣物也属于服装大类。

时装是指特定时期在款式上受欢迎的各种服装。在不同的历史阶段,流行服装的造型差别很大。在现代,几乎每个人都在某种程度上追逐时尚。如果年轻女士穿上她祖母年轻时所穿的衣服,就会显得怪异。不过,只有少数人的衣着会出现在高级时装杂志或时装T型台上。

区别基本型的成衣和时装并不简单,尤其是今天,时装设计师的设计灵感常来自于廉价面料和服装的功能性。例如,蓝色牛仔装早期是矿工和农夫的工作服。然而今天,即便是T恤衫,牛仔裤和运动装的简单搭配也受到时尚的影响。这一年,时尚牛仔裤流行细窄裤腿,下一年,也许又会流行宽松裤腿。

服装史学者通过多种方式来研究衣着的发展,如杂志和分类目录、绘画和照片、帽、鞋以及其他留存下来的东西。有关以往日常服装的可靠证据很难获得,因为大多数的出版物和图片资料都只关注富人们的时装。此外,过去留存下来的服装也并非是典型的日常装,例如博物馆里藏有大量的高级晚礼服,而普通工薪阶层妇女的日常装却非常少。早期男装被保存下来的甚至更少。图像资料,例如绘画、印刷品和照片,确实提供了日常着装的大量证据。这些资料表明尽管日常装通常不像时装变化那么快,但它也是不断变化的。

第二课 服装的功能

自史前开始,几乎所有的社会形态中,人类都穿着某种服装。有关人类服饰起源的理论也得到空前的发展。有些理论认为服装的起源在于功能性——保护身体,另外一些理论认为服装的出现是为了吸引异性——展示人体美。今天,现代学者认为除了上述两个功能之外,服装还是身份的标志及非言语沟通的方式。在传统社会,服装几乎可以作为一种语言,以表明一个人的年龄、性别、婚姻状况、出生地、信仰、社会地位和职业。在现代工业化的社会,服装不再受如此严格的限制,人们在选择所希望传递的信息方面有了更大的自由。即便如此,服装仍能展示穿着者的众多信息,诸如其个性、经济地位,甚或穿着者所出席场合的性质等。

一个社会的经济结构和文化、传统习俗以及生活方式也影响着人们的服饰。在许多社会,教规规范教徒们的行为,并规定精英阶层才能穿着某些特权服饰。即使在现代民主社会,服饰也代表着社会地位,贴有设计师标签的服饰价格相对昂贵,因此它具有了表现经济地位的功能。当服装作为制服时,明显地标明了社会身份,如警服和护士服以及神职人员或教徒所穿的特定服饰。

服装也从穿着时所处的场合衍生出含义。在许多文化中,新娘、新郎以及婚礼来宾穿着专门的服装来庆祝结婚典礼。在婚礼、毕业典礼以及葬礼等仪式上,按社会成员所认同的不成文规矩,穿着的服装一般是正装。服装也传达着参与某些活动属于何种类型的信号。某些类型的娱乐活动，特别是竞技体育，可能需要专门的服装。例如：足球、橄榄球、曲棍球球员穿着专门设计的配套的紧身球衫和裤子，以及相应的肩垫等。

现代社会由不同的社会群体组成，每个群体都有自己的信仰和行为方式。因此，每个群体就存在不同的服装亚文化。

第三课　纺织纤维

天然纤维

天然纤维来自动植物。除了家蚕丝，所有天然纤维的长度都较短，通过梳织和加捻形成一定强度的纱线，用于织造面料。家蚕丝从蚕茧中抽出可长达 2000 米，因此被认为是长纤维。

虽然纱线常被混纺或混织使用，许多服装仍完全是由丝、羊毛、棉或麻制成；与这些主要成分混合的其他人造纤维能增强产品的物理性能，并提高面料的美观程度。

使用化学过程可以增强或削弱纤维的某些特性，但纤维的结构并未因此发生改变。目前的基因研究正着手改变这种情况。天然纤维织物，尤其是棉织物，仍在消费市场上具有优势地位，尽管他们比人造纤维织物昂贵。天然纤维穿着舒适是因为它们与生俱来的吸湿性，美观的织物肌理，良好的染色性和手感。

人造纤维

人造纤维由化学方法生产，例如挤压液态化学物质通过小孔，在空气中或通过化学方法使之固化成纤维。他们可以是长纤维状，也可以切断成短纤维。人造纤维可以来自天然原料，如再生纤维；也可以来自化学或矿物原料，如合成纤维。

再生纤维素纤维是通过溶剂将木浆、棉花等天然产物溶解成液体，再重组加工成纺织纤维。合成纤维来自化学原料，主要成分是石油。

人造纤维始于复制天然纤维的特性。由于天然纤维优良的服用性能，因此，在当时的机械设备允许的情况下，早期的合成纤维都是效仿天然纤维的长度的。第一种再生纤维素纤维，是取材天然原料（木浆）而制成的人造纤维（粘胶人造丝），被称为“人造丝”，接着是醋酸纤维，再后来，天丝诞生了。面向服装行业的合成纤维的生产量已经超过了所有天然纤维的生产量。早期的尼龙、涤纶、腈纶特点明显，容易识别。直到最近，分析面料并对其纤维的性能做出某些推断才变得相对容易一点；而现在，纤维的识别又更加困难了。

纤维的结构影响着面料的外观、手感和舒适性。虽然纤维长度和表面形状是重要的，但纤维的内部结构也决定某个特定纤维的基本属性。纤维的形状决定面料的光泽：例如，蚕丝的长丝结构因为棱形而反光。改变吐丝器的喷孔会导致纤维截面的改变，可以模仿天然纤维的形状，也可以设计出新形状。这些截面可以是圆的、十字形的、三角形的、Y 形的或豆圆形的。纤维的结构形态也决定更多的机械性能：膨松度、硬度和吸湿性，例如，圆形截面的纤维具有抗弯曲性，Y 形截面纤维富于弹性，中空纤维相对于他们的体积而言较轻。不过，纱线的结构、面料的结构和后整理，必须明智地结合起来，才能满足审美与实用的市场需要。

第四课　面料

纱线和纤维通过编织、针织、交织或压烫等方式制成面料。很多情况下，只有当设计师看到面料才能确定未来设计的范围。面料给设计师留下的第一美感，是决定购买的重要指标之一。

面料的组织结构可以加强或削弱纱线的性能。几种主要的织造方法组合起来可以制造出令人眼花缭乱的外观;纺织品设计师常常依据稳定性和实用性对织物的外观和性能进行取舍。仔细检查面料可以分析出使用了哪些加工工艺。不同制造工艺的结合会混淆一般的分类。

面料的组织结构

织造面料的主要方法是针织和梭织。其他方法有:交织、刺绣和编结,但较少运用,这些方法多用于奢侈品或手工艺品的制作中。无纺面料有:毡布、各种类型的衬、聚氯乙烯衬布、熔合在聚氯乙烯底布上的绒织物等。

梭织面料

水平方向的纬纱和垂直方向的经纱交织而成的面料称为梭织面料。服装通常沿经向裁剪并制作完成。水平纬向或倾斜 45°方向(称为斜丝缕或斜向)的织物具有较高的伸缩性和悬垂性。

纸样裁片常常标注经向线以确保服装正确裁剪。

纱线的梭织种类很多,传统的织造法容易理解:平纹组织产生水平、竖条和格子的效果;斜纹组织产生斜向或人字形组织结构;提花组织可织造出复杂的花样;缎纹组织表面滑爽并富有光泽。不同种类的纱线加入纬纱或经纱会织造出不同的立体罗纹效果。起绒面料的组织结构中,处理经向(经绒)和纬向(纬绒)的纱线会产生不同的效果。组合不同的织造法,或者织造双层或双面面料都会产生许多罕见的组织结构。例如泡泡纱,就是对组织中的一组纱线不同程度皱缩处理而产生起泡效果的。马特拉塞凸纹布因加入经纱而产生绗缝的效果。当梭织结构中还涉及纱线的种类、色彩和印花时,组合的方式就更加复杂多样了。

针织面料

● **纬编面料**

纬编面料在纬编机上织造,纱线被上下移动的舌针牵引在面料横向上产生成排的联锁线环。

有些纬编机生产平面面料,有些生产圆形或筒形面料。纬编面料具有弹性,并随着纬编机器型号、纱线种类和纺纱张力而变化。成型的针织机直接织造出成型的衣片,但大多数针织产品是在平面或圆形纬编机上织造的。

各种各样的组织方法和图案:罗纹组织,衬垫组织,双罗纹组织,嵌花(也叫无虚线提花)组织和提花工艺等,为针织设计师提供了无穷的设计空间。纬编织法可以用于短期订单,因为没有梭织机或经编机那些复杂的经线控制设置。

● **经编面料**

经编机织造垂直的联锁线圈链。常用两股纱线来确保面料的稳定性。经编工艺发展迅速,机器生产很快,可使用人造细丝制造出大量的面料。这些面料尤其适合做女式内衣,也能织出网眼或花边的效果。

经编针织物的称谓容易混淆,特别是起绒织物,常见的有:抓毛布、毛圈绒、经绒、灯芯绒、圈圈绒等。经编面料还是很多粘合布和植绒面料的底布。使用中,根据用途或外观效果,决定起绒的一面是用于服装正面还是反面。

第五课　面料的性能

织物性能可分为三个方面：耐久性、舒适性和易护理性。

耐久性

耐久性是指所有影响一件服装使用寿命或功能的性能。包括面料的强度、保形性、回复性、抗磨损性和色牢度。

● **面料强度**

面料的强度能满足使用中的需要么？不同的纤维对外界的拉伸具有不同的抗拉强度。面料的强度也与面料的组织结构有关。质地紧密的梭织物、针织物通常较质地稀松的织物强度大。例如，帆布可用于缝制坐椅，而薄型网眼布只能用于缝制窗帘。

● **保形性**

织物穿洗后还能保持它的形状么？你的衣服穿久了，是不是膝部和肘部都变得松松垮垮的呢？面料在洗烫后，形状会走形，有些纤维在遇水受热时会收缩。

● **回复性**

面料能复原么？能在压皱后回复原来的形状么？悬挂能消除服装的折皱么？是不是必须熨烫才行呢？羊毛质地的地毯在一个沉重的家具下会被压扁，但是蒸汽熨烫后它又会回复原形。

● **抗磨损性**

衣物耐磨么？衣物与物体摩擦会产生磨损现象。领里与后颈的摩擦，腰侧夹带书本，这些都会导致磨损。有些布料会起球或在衣物表面形成若干细小的纤维球体。

● **色牢度**

色牢度是指面料上颜色的牢固程度，不会因为水洗，氯的漂白或日晒而褪色。但是，有些斜纹蓝牛仔水洗时会颜色变淡。马德拉斯狭条衬衫布，一种梭织方格布在水洗时颜色会渗开，以使格子变得柔和而模糊。

舒适性

舒适性是选择面料时另外一个需要考虑的因素。面料可能轻重适宜，质地耐久，易于护理，但穿起来却不舒服，也许会太热、太冷或太滑腻。面料的吸湿性、吸附性、透气性和拉伸性都会影响你身上服装的舒适性。

● **吸湿性**

这个性能是指面料对水分的吸收程度。有些纤维，例如棉和羊毛，吸湿性良好，其他的纤维，例如涤纶和尼龙，吸湿性较差。这就是为何大热天穿一件100%涤纶的衣服，你会感觉非常湿热的原因。你的汗液停留在皮肤表面，没有被织物吸收，这也是你用全棉毛圈毛巾比棉涤混纺毛巾可以更快擦干身体的原因。然而，某些特殊的后整理可以改进织物的吸湿性。

● **吸附性**

这个性能是指织物从人体上吸收水分并挥发掉的能力。有些纤维较强的吸附能力弥补了它们吸湿性的不足。烯烃，一种你将了解更多的面料，就具有吸附性。

● **透气性**

这是选择舒适面料时需要考虑的另外一个重要因素。它是指面料对水、气的通透性。有些面料经过专门的后整理具备防雨、防潮的功能，这些后整理也阻碍了人体表面水分的挥发,这就是当你穿橡胶鞋子和雨衣时常会感觉出汗的原因。厂家有时候为了弥补这个缺陷,会在防水服装的腋下增加汽眼作为透气孔。

● **拉伸性**

这个性能是指面料在人体上的"拉伸"和收缩能力。你的服装需要具备怎样的拉伸性,取决于你从事何种运动。泳衣、滑雪衫、运动装都需要更多的拉伸性。

易护理性

面料具有这种性能意味着服装或服饰易被护理:例如方便洗涤、干洗、熨烫、刷毛和折叠等所有面料的相关护理。

有些衣物比其他衣物要求更多的日常维护。当你选择衣物时,你应该选择那些符合你生活方式的。可洗性,防沾污性和抗皱性是影响衣物护理性的几个因素。

● **可洗性**

织物能被水洗么?或只能干洗?经过一个较长的时间,你干洗费可能已经累计超过了服装本身的价格。你有时间和空间去手洗并熨干各种服装么?虽然洗一件毛衫很容易,但是你会留出时间来做么?衣物是否缩小了1%或2%呢?如果是这样,那就影响服装的合体性了。

● **防沾污性**

面料能防污、耐污么?有些纤维会吸收污迹,但是专门的后整理可以帮助面料耐污和去污。地毯,室内软装饰,大衣和茄克,还有儿童服装常常需要这样专门的后整理。

● **抗皱性**

你每次洗完衣物都要熨烫吗?你是否每次穿衣前都对它进行熨烫?衣物短时间悬挂会消除折痕么?

纤维不同的性能影响面料的皱缩性。例如,涤纶抗皱性良好,但棉和人造丝则易折皱。特殊的后整理,例如耐久压烫,可以改善面料的抗皱性。

第六课　服装附件

服装附件或是具备功能性,或是具备装饰性,有的兼具两种功能。服装附件是服装结构中不可或缺的一部分。

服装附件必须与面料在服用性能和维护上保持一致。如何选择配饰件可根据面料的种类、日常护理以及款式结构来定。

附件:

- 衬布:针对服装某个部位加固的衬料,用于面料和里料之间。使得这个部位更加牢固或保暖。
- 里布:真丝、人造丝或棉制,女装的里布门幅通常为90～100厘米宽,男装里布门幅通常为140～150厘米宽。
- 粘合衬:用于增强或支撑领形、袖克夫或服装其他部分的硬衬。
- 帆布:用作衬布,最好的帆布是由亚麻制成,可与面料匹配。棉质帆布用于不重要部位的衬里,如袖克夫等。羊毛和马鬃制的帆布非常适合厚重服装。
- 加固麻布条:特别结实的,柔软的薄型麻布。用于加固口袋、扣眼和大衣下摆等位置。可用的有黑色、棕色和灰色。
- 西里西亚里子布:由缩绒的棉材料制成,用于大衣口袋的缝制。可选颜色很多,常用色有黑色、棕色、灰色和白色。

- 牵条:由亚麻制成,用于加固大衣的边缘、袋口等,1厘米宽。
- 衬垫:用于肩垫和袖山缝,有黑白两种颜色。
- 肩垫:某些服装使用,有各种尺寸和厚度可供选择,制作材料多种多样,有:羊毛、毡布、泡沫橡胶等。
- 格利萨特里子布:大衣袖子的里布,表面光滑,易穿脱。
- 缝纫机用棉、丝线:一种光滑的细棉、丝线,S捻,有各种规格可供选择,有线轴线团和纺锤线团两种。
- 粗缝棉线:一种粗糙的白色棉线,易拉断,有线轴线团和纺锤线团两种。
- 丝线团:一种细而坚韧的手工用丝线,主要有三种颜色:蓝黑、棕褐色和灰色。
- 拉链:用于服装、包件的紧固件。由两排联锁的金属牙或塑料牙组成,并有滑块上下移动开合紧固件。
- 钮扣:一块扁平的圆型配件,通常由塑料或其他材料制成,通过扣入钮孔或线圈来连接服装的两部分。
- 风钩:通过将一个小挂钩插入一个金属环或线环来连接服装。
- 按扣:由一对圆形紧固件构成,按压即闭合,拉开即打开。
- 缝纫线:两组或更多组纤维合捻而成的细线,用于缝纫、编织。
- 亚麻线:有线团、线轴两种。线团线较强韧,用于钉钮扣和其他加固缝纫工艺,也用于开钮洞。
- 橡筋:橡筋是一种带状或线形的橡胶制品或类似的弹性材料制品。在面料或织带中加入橡筋可以增强合体性。
- 标签:由纸、布料或塑料制作的服装说明或标识。

饰件:

- 线形饰边:来自服装上的线条和布边,装饰布边和缝线即形成线形饰边。
- 窄条饰边。
- 缎带:缎带是边缘整齐的窄条梭织带,宽度从0.3厘米到几厘米不等。与其他织物相似,品质多样。
- 边饰:边饰是指一系列直条的、弧形的、流苏的或穗状的镶边和滚条的统称。
- 蕾丝:蕾丝是由很多线缠绕在一起形成网眼状的饰带或面料,图案复杂,质地轻薄透气或厚实严密。

第七课　肌理、图案和色彩

肌理

肌理,是影响织物手感的视觉因素,是面料结构、纤维、纱线和后整理的组合结果。

在人台上立体裁剪制作服装,会帮助制作者了解布料不同造型后的表现。将平面的布料转化成立体的服装,就如同陶工用泥土制作陶盘、碗等陶器一样,是成衣设计师必不可少的训练。

有光泽的面料反射光线,并在视觉上产生膨胀感,尤其是紧身服装。光泽暗淡的面料反射光线较弱,能够柔和地烘托出穿着者的面部。真丝绉就是一种典型的低光泽面料。无光泽面料就是没有光泽,不会突显相应的人体部位,当无光泽面料是中性色或深色时,尤其适合用于伪装。

挺括的、吸湿性良好的面料因为吸汗和透气性适合制作夏装，例如棉和亚麻。柔软、膨松的海绵织物因为面料的厚度而保留了人体的热量，从而给人温暖的感觉。厚重的羊毛织物和绗缝面料适合做冬天的外衣，因为它们可以有效地阻止热量的散失。

观察并体验各种面料，了解纺织品内在的特性是如何形成服装的实用性和多样性的。

面料图案

面料的图案由色彩、线条、形状和空间构成。图案种类非常丰富：条纹、格子、几何形、植物、风景、边纹，以及其他多种类型。图案有大有小，有规则的或不规则的，有深有浅，有分散的也有集中的，有暗淡的也有醒目的，这一切都会影响服装着装后的效果。

正如面料肌理一样，面料图案也可产生视错。色彩柔和的小图案通常收缩服装的体积感，大型图案则会增加体积感，间隔大的图案也会让你看起来体型更大，大弧线的印花给人以膨胀的圆球感和体积感。

遵循基本的线条和视错原理来选择条纹或格子面料。如果你不能清楚地辨别主要的条纹，试着眯起眼睛观察，当你眯起眼睛去看时，最先跃入眼帘的就是主干条纹。这些条纹的位置是重要的，你将在服装上如何布置这些条纹呢？垂直放置？水平放置？还是对角线排列呢？

服装上的条子按照什么方向放呢？例如，一条醒目的水平线横贯臀围或腰围，这两个部位看起来都变粗了。条纹遇到接缝要怎样处理呢？它们会在接缝处出现山字形或形成尖角吗？

根据自己的体型选择大小适中或比例相称的印花、条纹和格子布。小花型适合小号体型和中等体型的人，大号体型的人穿着不合适。另一方面，大花型适合中等体型和高挑身材的人，小号体型会淹没在大花型中。

色彩

人们已发现，色彩是服装吸引顾客的第一要素。色彩是最基本的时尚元素。常常是设计师每季需要做的第一个策划内容。

- **色相**

一个颜色（如红、蓝）与另一个颜色（黄）之间的区别。

- **纯度**

某种颜色的饱和度或亮度——明度和灰度之间的区别。或者说，色彩中灰色的含量。例如，一个灰蓝和深灰蓝，色相与明度相同，但是灰度不同。

- **明度**

具有同种色相和纯度的亮色（例如，亮绿）与暗色（例如，灰绿）之间的区别。这取决于加入的纯白与纯黑有多少。

第八课　设计原理

比例

简单地说，比例就是服装上局部与整体的关系。人体由许多不同形状的部分组成。设计师必须调整服装的各个部位，通过强调人体的自然形状或创造出一个全新的造型，来美化人体的体型。设计师会观察对象的空间，从高度和宽度方向上进行划分，去构建一个美观的外形。经典、自然的腰线比例是：上身∶下身约为3∶5。

然而，其他比例也同样被采用。20世纪90年代中期青少年装市场上就流行一种“美少女装”，较短的上衣和较长的裙子组合搭配，使少女显得更加年轻和高挑，高腰节夸大了少女的人体比例，相对较宽的肩膀形成的楔形则消除了造型的粗壮感。

平衡

设计师在水平和垂直方向设计服装的分割线，为了使服装外观具有吸引力，必须对每一个分割位置进行一定程度的刻画和强调。任何部位有太多或太少的设计都会显得不平衡。

统一

统一意味着所有的设计元素搭配和谐，彼此不产生冲突。例如，一件茄克是偏门襟的，对应的裙子就不能是居中的门襟。所有的元素看上去都应该是安排好的，不能出错。

重点

简单地说，重点就是服装上的焦点、中心。如同每幅画都有一个有趣的中心一样。重点有可能是某种面料、某种色彩、某个细节或某个配饰，但它是顾客留意这件服装的主要原因。

轮廓

轮廓是整个服装的外部形状，在某个特定时期，是所有服装最共通的元素。当你在电影中看到垫肩和严肃的军服套装轮廓时，你知道它是 20 世纪 40 年代的作品。

紧身针织上装和大喇叭裙的装束，是典型的 20 世纪 50 年代的造型。造型轮廓的变化是缓慢的，这也是不同设计师的作品容易雷同的少数原因之一。例如，当流行宽大的轮廓造型时，几乎每一款服装都是尺码超大的。

线条

服装的线条包括服装的分割线和装饰线。线条通常会营造出一种“视错效果”：更长、更高或更苗条。例如，公主线会让人显得更加纤细。但是有时候，线条被用来制造重量上的错觉。例如，一个瘦弱的女性穿着一件无吊带衫会非常漂亮，因为肩膀显得更宽了。强烈的不对称线条会制造一种醒目的视觉效果。

第九课　时装设计师

时装设计师通常都专门设计一类服装：例如男装、女装或童装。在每一种类别中，他们的目标常常锁定某个特定的年龄段或生活方式，例如男装会锁定消费对象是 25～40 岁男性。有些时装设计师负责几种不同的产品，特别是那些在小公司的设计师，或大企业中的资深设计师。设计师们的分工种类有：

- 高级订制
- 休闲装
- 运动装
- 针织衫
- 弹力服装
- 内衣和睡衣
- 泳装
- 特定场合服装
- 俱乐部服装
- 晚礼服
- 晚服

有时候，在大公司里，设计师仅需设计一类服装，如女茄克外套。例如玛莎百货的供应商们，就只提供大量的一种款式的各种尺码/号型。这有助于设计师获得细节设计经验，但是，有些设计师可能会感觉在有限的产品设计空间中，设计受到了限制。除了服装，配饰是重要的时尚产品，设计师也被要求设计帽子和鞋子。

时装设计师有不同的职责，无论是身处哪一类市场，大多数设计师的工作都包含以下内容：

- 流行趋势研究；
- 产品定位和市场比较；
- 采购面、辅料；
- 设计；
- 产品范围介绍；
- 商业策划会；

根据公司的安排，有的设计师还需要做以下工作：

- 编制设计规格表；
- 裁剪纸样；
- 试衣；
- 为时装展示准备服装。

设计

设计中必须考虑潜在消费者的生活方式。设计师根据以下因素构思理念进行设计：

- 造型轮廓
- 设计细节
- 面料
- 色彩
- 图案
- 配饰与紧固件等

设计策划

时装设计师在计划商业生产之前就已经构思了大量的初步设计理念，最恰当的设计思想将进一步深化并被绘制成一系列的草图，以观察可能出现的效果。例如，一旦设计师为一款茄克选定某种轮廓，在得到确认样之前，不同的领形、口袋和缝缉线会被应用到设计中，进行尝试。时装设计师运用他们所研究的流行元素为目标市场设计合适的产品。设计师可能会自己决定设计方向，或向资深设计师咨询，寻求有关他们设计概念的商业性的建议。无论他们服务于哪个市场，设计中他们都必须考虑以下因素：

- 消费者审美取向；
- 零售商的价格范围；
- 面料与成衣的技术性能。

设计师的创意程度受这些变量约束，这有助于发现商业解决设计主旨的方法：这个产品将成为奢侈品设计师发布会的一部分，还是价值引导的大众市场的成衣。其他因素可能会在某些公司的设计理念中占有重要地位：服装业和纺织业所引发的道德和环保问题，消耗大量的地球资源等，均是日益关注的焦点。为了迎合现代消费者的关注焦点，一些品牌在他们的产品中使用有机纤维，这种“绿色”的做法也被遵守公平贸易和生态政策的成衣公司采用。

很多设计师试图拓展他们有限的创造领域，这被认为是与创造的个性并行的优点。设计师常常因为过于专注于效果图的绘制，缺乏工艺技术而在行业中被指责。那些在工艺和资金允许的范围内，能不断产生创新概念的设计师被认为是服装业的宝贵人才。

第十课　时装的演变

标新立异者最先穿着新潮服装，旨在有别于普及的大众时装。通常，大多数别致的服装新款都来自欧洲高价位设计师之手。新潮服装的淘汰率较高。有些新潮服装是来源于“街头服饰”，由极具个性的时尚达人组合、设计、搭配而成。

服装专业人士常常寻找新的流行趋势。当某一种风格开始“风行”或出售给那些时尚先锋们，仿真复制也开始了。

在流行的第二阶段，流行创新阶段，原创风格被各种面料制作成不同的版本。时尚先锋常常是社会名流，电影明星以及出现在媒体中接受采访的人们。这阶段，广告在各类光鲜的时尚杂志上频繁出现。

接下来进入流行的接受阶段。新的趋势占据了高级专柜和专卖店的橱窗、各类图像广告的重要位置。大量零售商的买手们和制造商开始以较低价格成本复制这些款式。在接受阶段的顶峰时期，款式被大量仿制，使用的面料和颜色也缤纷多彩，并且频繁出现在邮购产品目录册的最新款式中。

流行过程中，当基本款被大多数人接受时形成经典服装。通常，经典服装与其他新鲜元素组合可以让他们重新焕发生机。经典的服装可以由各种各样的面料制作，以更广的价格范围赢得大众的需求。褶裤就是这样的一个例子，稍微调整一下，就又能表现一个时期的流行。例如，裤腿或许是锥形的或者宽脚口的，有折边或没有折边，但是裤子本身却保持常年的流行。

当出现越来越多商家自制的复制品时，意味着这个款式已经进入了推广阶段。制造商的预算部门会策划使用廉价面料来生产，以降低成本。当低价面料出售给成衣商或折扣商，流行开始衰退，这些款式常出现在促销或批发市场。

流行的最后阶段是流行周期的淘汰阶段。这阶段中流行的款式已经在低价复制品中泛滥，大众逐渐抛弃了她，款式也因为过季而卖不动了。

但是，先等一等，几乎所有的时尚潮流都会经历整个周期，然后被新一代的产品重新演绎。时装创新者常常从全新的视角观察，将某个时期的流行款式重新组合而形成新的风格，等市场上这种款式已经饱和时，再引导另一个几十年的流行。

商业设计师们分析他们的顾客，把他们放在流行周期的适当位置来研究。设计师们将流行周期看成一个整体，分析新的趋势，并根据顾客的生活方式和流行周期选择相应的设计种类、色彩和造型。

第十一课　流行预测机构

时装发布会

发布会报告是设计师发布会上最立即可用的资源，但是它们是昂贵的，这些报告提供新设计的细节效果图，包括色彩和面料选择，这些都是来自某个最新的设计师时装展。许多公司会在设计师发布会后两周后得到这些资料。

流行趋势发布

有几家机构编写流行趋势：Tobe Report 和 Block Note，此外，Mode 这家公司提供的网页，也包含所有可用的流行趋势。流行趋势报告类似于时装发布，预测机构为时装买手和设计师提供各类效果图、照片和面料/色彩小样，以协助商业决策。流行趋势由独立经营的公司编写，不是来自任何一个设计师发布会。

与流行趋势相似，色彩机构提供有关未来一季的最新流行色的预测信息。专门研究色彩预测的公司有：国际色彩权威、色彩盒子以及色彩概念。色彩体系用于色彩的流行预测。Pantone 和 SCOTDIC 提供了两套色彩体系，这两家公司均留有染整面料过去曾使用过的所有色彩的记录。

网站传媒

时装预测网站和提供简单的流行讯息给设计师与买手的网站是不同的。时装预测网站通常需要付费预定才能访问。这种网页面向买手，这是他们收费的原因，也有很多商业网站更适合普通消费者，买手也会去看。

这些可用的网络资源提供的信息，无论收费与否，都能帮助时装从业者们更好地策划设计生产，或经营合适的产品。

商业出版物

商业出版物可能是报纸、杂志或者由专业的时装行业人士编写的期刊。

时装从业人员阅读商业出版物是相当重要的，可以了解最新的趋势、行业合并以及其他的变化。阅读时装杂志也很重要，有时候被称为编刊或大众期刊。

除了商业出版物之外，商业协会也可能提供流行讯息。一个商业协会就是一个由商家和厂家组成的行业组织，去促进商贸交流或产业发展或接受统一的标准规范。

电影、音乐和电视

许多优秀的流行预测者会关注电影电视明星的衣着，除了给设计师带来灵感之外，明星们的穿着还有助于预测流行趋势。名流常常将顶尖时尚展示给大众和高级成衣市场，但这并不是真正流行趋势的开始。

街头时尚

最后，街头时尚或人们的日常装，也会给流行预测提供信号。通过简单观察的方式，例如研究人们在购物中心、音乐会或其他活动中的着装也可以发现新的流行。

第十二课　时装设计过程

时装设计过程是一个漫长的过程，根据公司的类型存在很多差异。一般的设计步骤包括：

1.市场调查、系列企划；

2.构思设计理念；

3.系列设计；

4.生产计划和加工；

5.系列营销策划；

市场调查、系列企划

这是设计师研究面料趋势、色彩趋势、研究过去销售业绩和制定生产目标的阶段。有些公司将生产自己的产品，称为私营品牌。其他公司只买制造商的设计。当零售商有自己的设计师为他们设计时，出现了家庭作坊或私营品牌。设计师们将参阅时尚杂志，浏览色彩预测，甚至观察名流们的穿着，去了解最新热门的流行将会是什么。这第一阶段为后面的工作打下基础。

构思设计理念

接下来的设计阶段即是形成设计理念，在生产之前，这个理念实际上会卖给公司内部的买家。这个阶段，设计师将要绘制效果图，收集面料样本，设计一个系列，通常一共八套款式。这个系列策划然后被送至公司内部的决策者，并决定是否进入下一个阶段。

许多工厂在系列的策划中采用计算机辅助系统，CAD 系统也支持纸样的绘制和裁剪。服装制造商也具备了采用 CAD 进行产品设计、生产船运和销售管理的经验。

系列设计

当设计被接受，可以进一步深入后，设计师或打版师将为这些服装制作样板。样板的制作有两种途径，第一种是平面打样，在厚纸上手工裁切出样板；第二种是立体裁剪，通过在人体或人台上铺放面料，用大头针假缝来制造样板。

纸样打出来后，通常用细布将样品制作出来。细布，是一种廉价的面料，适合做样衣。然后，如果设计师满意样衣效果，一个样衣缝纫工就用实际面料把款式做出来。样衣缝纫工通常是公司内部的，有时候他们也要为海外制造商工作。在这个阶段，产品的价格也确定下来。

生产计划和加工

由于全球经济化的扩展，许多加工是在海外完成的。基于此，设计师需要制作工艺规格文件包，有时简称工艺单袋。

一个工艺单袋包括所有有关服装细节的图表和工艺说明。细节说明：例如，使用的闭合件类型或领子的类型等。工艺单另一个内容是对服装缝制工艺、方法、缝线位置等的专业说明。有时候，制造商会邮寄一款根据工艺单制作的样衣给客户，来确认正确的放松量和预期效果，这个过程中一个重要部分就是与海外客户的有效沟通，一旦工艺单袋寄出，制造商就开始了裁剪、缝制，直到成品完成，随后货物按照要求被装运。

系列营销策划

系列营销也被认为是市场经销或系列产品的实际销售。营销的定义是顾客获得实际产品的方式。营销包括和海外客户的合作过程，海外船运货物，以及在商店里陈列货物。

第十三课　模特

模特是时装界中不可或缺的一部分。没有模特，服装就没法试穿、评估以及展示给买手们。公众只看到世界时尚之都秀展里的超模，但是，模特的种类等级却有很多。虽然模特有男有女，实际上，女模才是时尚过程中最出色的副产品。

女模特

尽管厌食症和过于年轻化备受争议，模特的苗条、优美比例的体型仍然是最重要的。她不仅要苗条和高挑，而且还必须拥有完美的肩形、纤细的腰围、窄小的臀部和修长的腿型。回顾过去 60 年的时装图例，你会发现这些体形特点都是为了让服装看上去更好。这些在模特身上表现完美的服装，穿在体型比例较差的人身上，也想保持同样的形状是不现实的。

模特可以是走台模特，也可以是摄影模特。常常一个优秀的走台模特不是摄影模特，摄影模特做走台模特高度又不够。

各种等级的模特都是需要的。因此，为了这个需要，许多城市都有模特经纪所。过去，高级时装屋雇佣室内模特，摆出一个或两个固定姿势去展示所有的样衣，并举行小型室内展示。他们也雇佣几个其他类型的女模为每季发布会展示。这些，都是女模获得经验的途径。

顶级的模特经纪所，例如纽约的 Models One 或 Elite 仅仅从事那些可能会出现在国际天台上或顶级时装杂志或大型广告活动中的女模特们的经纪事务。许多模特从事更加平民化的待售品目录的工作，以获取高额报酬。

其他等级的模特用于贸易展，例如车展、船舶展和许多时尚产品展，例如，英国伯明翰的布料展。

男模特

从时装的视角来看，男模在许多方面没有女模重要。只要他足够高，相貌英俊，可以走一条直线，极有可能就会在街头被发现，登上天台走猫步去了。拍摄工作，就是另外一番景况，男模必须相当英俊、上像，有这些因素才能有助卖出产品。这个领域，男人可以达到超模的地位，尤其在激烈的广告战中。例如，1995 年，沃纳·施耐尔就是在雨果波士品牌中成名的。

第十四课　配饰设计

服饰配饰是对服装的补充和装饰，种类有：珠宝、手套、手袋、帽子、腰带、围巾、手表、太阳镜、别针、丝袜、领结、护腿、袜套、领带、吊带及裤袜等。

配饰给服装增添了色彩感、时尚感和品味，形成了某种特定的外观风格。但是，他们也具备实用功能。手袋可以用于携带物品，恶劣的气候下帽子可以保护面部，手套可以保暖。

许多配件由服装设计公司制作，但是，越来越多的个人，通过设计生产自己的配饰商标来创建自己的品牌。

配饰可能被视为宗教或文化起源的外部视觉符号：耶稣十字架、伊斯兰头巾、牧师的无檐便帽、穆斯林缠头巾都是常见的例子。配饰上的设计师标签被视为是社会地位的标识。

- 戒指品质多样，男女均可佩戴。金属、塑料、木材、骨头、玻璃、宝石或其他材料均可以制作成戒指。戒指可能会被装置某种“石头”，通常是贵重的宝石或半宝石，例如红宝石、蓝宝石或翡翠，也可以使用其他任何材料。
- 胸针是别附于服装上的装饰类首饰，一般由金属制成，常用材质有金、银等，有时候也用铜或其他材料。
- 手镯是手腕上的环状首饰。手镯可由皮革、布料、麻绳、塑料或金属制成，有时候镶嵌宝石、木头或贝壳。
- 耳环是穿戴在耳垂或外耳廓其它位置的耳洞上的首饰。耳环男女均可使用。耳环可由任何材料制成，包括金属、塑料、玻璃、宝石、串珠以及其他材料等。
- 腰带是一种柔软可调节的带子，通常由皮革或结实的布料制成，环绕腰部。腰带用于固定裤子和服装的某些部位，可用于造型和装饰需要。
- 手袋或手提包通常设计的很时尚，多为女性携带，用于放置钱包、钥匙、纸巾、化妆品、发梳、手机或私人数码产品、女性卫生用品等。
- 手套是一种包裹手部的配饰，手套根据每一个指头设计了独立的指套或指口。只有一个开口，没有分开的指套的手套称为“无指手套”。无指手套中只有一个大开口，没有单独的指头开口的手套称为“防护手套 ”。包裹整个手掌，但不具有单独的指头开口或指套的手套被称为“连指手套”。
- 帽子是一种头部配饰。它的使用可能是为了抵御外界不良气候，或是宗教原因，或是安全需要，或充当时尚配件。

- 围巾是戴在头上，或围在头上或颈部的一块布，它的使用可能是为了保暖、防尘、时尚或宗教原因。
- 手表是佩戴在身体上的时钟。现在手表一般指的是用皮带或手镯佩戴在手腕上的腕表。
- 太阳镜是保护眼睛的装置，用于保护瞳孔防止被强光和紫外线损伤。
- 别针是用于将物体或材料固定在一起的一种工具。通常由钢制成，有时也用铜或黄铜。将材料加工成一根细金属线，锐化末端，添加别针头即成。
- 长袜有时称为长筒袜，是覆盖脚和腿的下部，具有不同程度的弹性紧身服装。长袜有不同的颜色和透明度。
- 暖腿用于保护小腿，类似于袜子，但更加厚实，一般没有足部。暖腿最早是芭蕾舞者和其他古典舞者穿用的。
- 袜套是由好几块合体部件组成，用于保护腿部。最初的袜套是由两个分开的裤腿组成。
- 领带（或领结）是围绕在颈部或肩部的长条布，位于衬衫领下方，咽喉处打结。领结包括蝴蝶结、领巾式领带、饰扣式领带和夹式领带。
- 吊带或背带是穿戴在肩部用于固定裤子的布带或皮带。
- 裤袜是一种下装，大多数的长度是从腰部至足部长，或多或少有点紧身。
- 项链是一种围绕在脖子上的珠宝装饰。常由附加的吊坠或挂件的金属珠宝链制成。

第十五课　服装纸样设计原理

基本原理适用于许多纸样裁片，这些原理应该在打样前就要考虑到。

接缝

衣片可以沿纵向、横向、斜向直裁开，也可以用弧线分割。当裁开的衣片缝合时，基本形状不会发生变化。但是服装上出现了拼缝线。省道也可被转移至拼缝线中，结果是服装造型未变，但是省道消失了。

外形

服装可以是紧身的、半紧身或宽松的，这取决于如何处理原型。例如，可以通过加入额外的松量来放大尺寸，设计褶裥来收身，增加塔克和碎褶来形成喇叭造型，或仅仅增加下摆的围度形成锥形造型。

增加零部件

当增加口袋、腰褶、嵌条、门襟等部件时，需要仔细考虑设计的平衡。

人体的运动

对大身纸样的进一步加工即是袖片的裁剪，在这一步的操作中需要注意人体的运动需要，这里只需要在宽度上留意即可，宽松的服装可以采用非常简单的外形。

完美造型

优美的线条和形状是必需的。当裁剪复杂的纸样时，基本原型会舍弃某些细部结构，或另外增加小块衣片，如何取舍取决于个人的技术和经验，这就是裁剪顶级设计师的纸样非常困难的原因。

裁剪单独的服装给予设计师更多的自由，他们不再受制于有限的价格以及成衣生产中的面料。

生产商要求大生产的纸样必须有缝份。有些生产商会要求设计师修改那些根据毛样原型绘制的纸样。

缝份宽度随着加工和服装种类而变化：

- 基本的缝份，例如侧缝线、款式线，1 厘米到 1.5 厘米；
- 止口缝线，例如领口，克夫，0.5 厘米；
- 下摆卷缝根据形状和大烫工艺来定，1 到 5 厘米；
- 装饰线通常要求更多的缝份量。

易磨损的面料要求在挂面和领子上有更宽的翻折量，缝份的宽度必须用直线和刀眼在纸样上标示清楚。

折叠线不需要缝份。

添加至纸样的缝份标记准确和清晰很重要。

样衣纸样

这阶段必须为纸样增加缝份，可以直接标记在白坯布上。

工业纸样

净缝线不标在(工业)纸样上。缝份通常会在生产工艺单中标明，并且只有可调节的缝份会被刀眼标记。

为了使服装制作正确，下列的注明必须标在纸样上：

- 每片裁片的名称；
- 前后中心线；
- 裁剪的衣片数量；
- 折叠位置；
- 对位记号：帮助确定纸样在正确位置拼合；
- 缝份：沿着纸样的外围，在每条缝线的末端使用线条或刀眼标记出缝份。如果纸样是净样(不含缝份)，需要将此信息清楚地绘制在纸样上；
- 结构线：这包括省道、钮洞、袋位、塔克、褶裥线、装饰线迹等。结构线需要标示在纸样上或钻孔显示；
- 经向线：为了获得理想的效果，你必须理解在面料上沿着正确丝缕方向排料的原理。使用一个箭头表示丝缕方向，在纸样裁剪之前将这个方向标记在纸样上，一旦纸样裁开，在各个裁片上要想找到正确的经向线就困难了。
- 纸样尺寸；
- 款号。

第十六课　平面纸样裁剪

平面裁剪纸样是时装业常用的方法。根据零售商或品牌商的标准尺码来制作纸样，通常女装使用的是英国码 12 号(欧板 38/40 或美板 8/10 码)。

基础原型首先经过调整来获得款式所需的放松量和相关的结构细节，并通过透明纸或描线轮转移到卡纸上。为了提高打样速度和从节约角度考虑，学生常常被建议使用 1∶5 的小样，实际上，工厂一直使用实样。初期草稿上原型边线就是服装的缝合线，因此，一个工业纸样，需要在所有边线外追加 0.5～1.5 厘米缝份，确保服装能被缝制。

缝份的尺寸取决于缝纫设备和缝纫面料，由靠近纸样边角的刀眼表示。纸样裁剪师需要麻利而精确地处理细节。他们精确测量尺寸，如果没有采用正确的缝份，服装的合体性和生产质量将会受到不良影响，并导致高返工率和滞销。纸样裁剪师需要全面了解服装的结构。

原型的许多方面能被修改来获得所需的造型，包括修改省道来调节合体度和造型，增加拼缝设计细节，为上装增加立领或翻领，利用褶、裥或皱折来增加裙子的丰满度，修改下摆的长短等。当纸样上的造型线和设计细节完成时，纸样需要增添挂面和里子，这些将有助于服装的完整性和结构，这可以根据是否需要来增加。

纸样需要下列的标注：

- 服装参考编号及款式名称；
- 服装尺寸；
- 前后中心线；
- 经向线；
- 折叠线；
- 对位点（刀眼标记，确保前后片对位缝制）；
- 裁片数量（确保不要漏裁）；
- 裁片名称，因为有些裁片看起来非常相似；
- 结构线，例如省道、钮洞和口袋位置。

面料的经向线与布边平行（布边是面料固定在纺织机上的那部分）。纸样上，经向线表明了丝缕的方向，标示了裁片在面料上放置的位置。设计师可能会故意安排服装的一部分斜裁，例如腰带部分，利用面料斜丝绺提供的拉伸性。

对于一个优秀的裁剪师而言，把各个纸样组合成一个型号是重要的。这样，即使裁剪师不在公司，同事也可以容易地找到需要的纸样

由于针织物面料的弹性太大，纸样的裁剪趋向于相对简单的形状。“裁剪成形”的针织物由若干块“坯布”缝制而成。“坯布”事先被裁剪成所需的形状，用包缝机包缝连接成形，随后服装的领口，克夫和下摆也被缝制起来，常常使用针织罗纹组织。全成形工艺是一个传统并高品质生产针织服装的方法，所有的衣片均直接编织成形，不需要裁剪，因此浪费率很低。

第十七课　立体裁剪

面料常常是设计灵感的来源，但是，正是立裁师们高超的技艺，才将不成形的衣片制作成精美的服装。

多数服装的第一件样衣是由坯布立体裁剪制成的。因为坯布是一种廉价的面料。理想状态是使用设计指定的面料或肌理和重量都相似的替代品进行立体裁剪。尽管就制作者而言材料太贵，但是对于私人订单是适合的。在立裁完成时，服装从人台或模特上取下之前，将就造型线位置、比例、平衡和合体性等方面进行评估。

在评估后，别针从立裁作品上取走，标记好的造型线和线迹被描实。描实立裁服装要求将坯布上所有的标识描成流畅的直线形和弧线形的结构线。尺寸将与尺寸表上的数据进行比较。如果想要获得准确的样板形状，这些步骤是必不可少的。随后样板被缝制并放在模型上，或由模特试穿，进行合体性测试，也可以描到纸上进行裁剪，并使用成衣面料缝制后，进行合体性的试穿测试。

初学者常常会惊讶地发现，人台上别缝的完美造型，当使用成衣面料裁剪并缝制成形时，仍然存在不合体的现象，这是因为下列两个原因：

1.成衣面料在肌理和重量上有别于假缝面料，这导致服装外形悬垂感不同，因此出现了不合体现象；

2.对坯布上样板的不准确标识和描实，当转移到纸样上时，也导致了不合体现象。

一开始，立裁设计师们都会努力在人台上设计制作美观的作品，但是，立体裁剪并不简单，每一个成功的设计师都知道，只有艰苦的工作，才能获得理想的效果。立体裁剪依靠对放置于人台上的面料的经纬向的控制，来获得理想的设计效果和协调的搭配。引导制作者正确立裁的是对服装结构原理的理解和掌握。最后，立裁师必须具有面料性能的知识，并且有能力选择与设计相配的面料。合体性的鉴别和解决办法是一个持续学习积累的过程。要想进步，记住关键是坚持，以及对立裁服装的热爱。

为零售市场设计的服装要求工业样板，服装在投放生产前，通过几个测试板来确定满意的合体性和纸样。

为私人客户制作度身定制服装的设计师们，常常用成衣面料进行立体裁剪。有时，立体裁剪完成后，服装被取下，假缝试穿和调整修改后，服装被缝制成衣。不需要纸样绘制，整个过程就结束了，因为它不会再被制作了。

第十八课　服装结构术语与标识

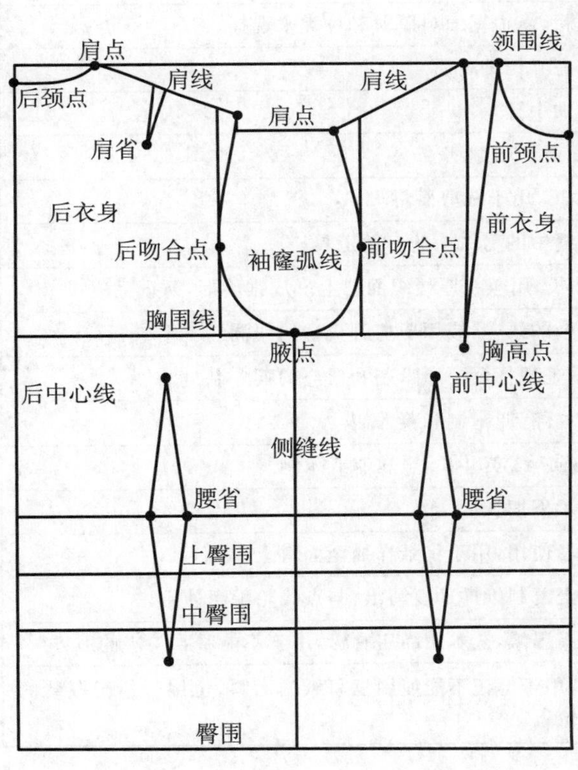

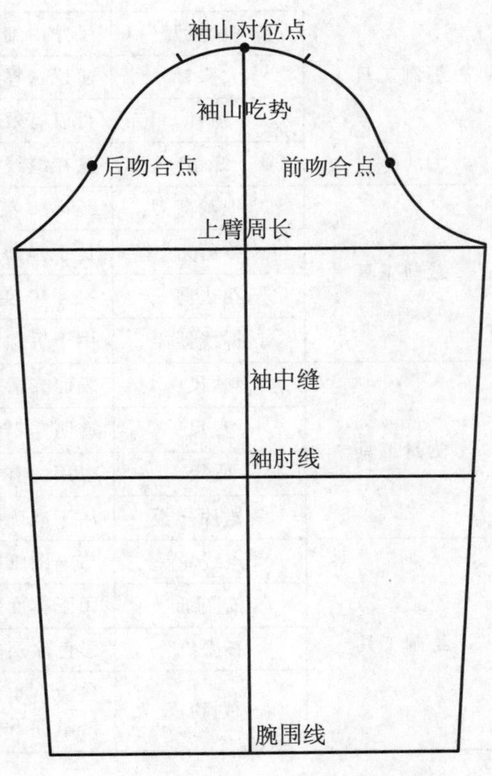

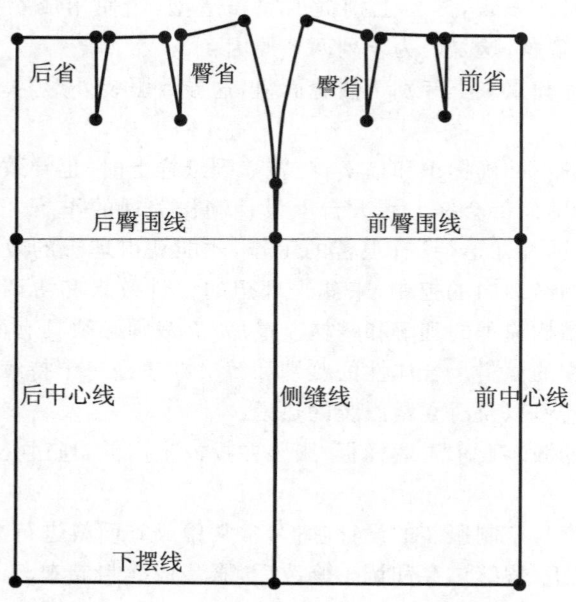

第十九课　缝纫设备

缝制工具	手缝针	尺寸由粗(1号)至细(10号)
	机缝针	尺寸由细至粗,需根据面料厚薄和种类来选择
	大头针	建议选择光亮不锈钢材质
	顶针	可以有效帮助手缝
	蜂蜡	涂在缝线上,有利于缝纫
裁剪工具	弯把裁剪刀	8至10英寸长,用于裁剪面料
	纱剪刀	是手缝和机缝中的剪线头的有效工具
	花边剪	9至10英寸长,用于在厚梭织面料上形成锯齿形,防止毛边
	拆线器	用于拆除多余的线迹,或用于剪开机缝的钮洞
测量工具	软尺	表面光滑,清楚地标示了厘米制和对应的英寸制
	直尺	透明塑料制,标有厘米制和英寸制
	码尺	可用的有36或45英寸长,金属制或木制
	裁裙片样板	用于准确标示下摆长度
绘制工具	点线轮	与描图纸配套使用,用于将纸样描至面料上
	描图纸	用于在面料上复制纸样的复写纸,与点线轮配套使用
	石墨纸	彩色涂层的石墨纸,艺术用品店有售,用于不能看到碳迹的地方
	划粉	蜡或石头制,用于标记不能使用复写纸的面料,也用于标记服装的松量和下摆位置

续表

缝纫工具配件	嵌线压脚	左右两侧均可使用，缝制效果均匀、紧密，用于包缝线迹和安装拉链
	滚边压脚	用于加工斜丝缕的滚边带
	碎褶压脚	用于制作均匀的、定型的多层抽褶
	双边固定轮压脚	适合缝制皮革，合成人造麂皮和聚乙烯基织物
	暗拉链压脚	用于缝制隐性拉链
	狭卷边压脚	用于缝制下摆的卷边机
各种其他工具	翻带器	用于将斜条缝制成“细肩带”和窄带
	锥子	用于牵引缎带或橡筋穿过抽带管
	人台	放置在一个可调节底座上的专门的人体模型
熨烫工具	蒸汽电熨斗	可调节温度的大型熨斗
	熨衣板	加垫的熨板，固定在地板上
	烫袖板	加垫的小型熨板，用于熨烫袖片和细窄部位
	袖烫垫	加垫的馒头，用于熨烫极难接触部位的线缝
	馒头烫垫	用于熨烫和定型曲线部位
	熨烫板或马凳	未加垫的设备，有助于熨烫重点位置
	针毯烫垫	平绒、丝绒和其他起绒及毛皮面料的熨烫必备设备
	手套式烫垫	套在手上使用，用于小面积熨烫，避免接触到服装的其他部位
	熨烫垫布	由质地较厚的斜纹棉和羊毛织物(制成)，防止面料正面熨烫时出现极光和烫焦
其他	锁式线迹缝纫机	用于缝制直线性线迹
	曲折缝缝纫机	用于贴饰花边，加入橡筋，为经编织物提供一个装饰性工艺
	包缝机	用于裁切、缝制、包缝面料
	安全线迹包缝机	使用 4 卷线为面料提供额外的包缝线迹

第二十课　缝制过程介绍

缝纫主要用于缝制服装和家居用品。事实上，缝纫在服装制作中是重要的环节之一。许多工业缝纫都是由工业缝纫机完成的。服装的裁片一般先被固定，或者假缝起来，随后缝纫机上复杂的部件会牵引缝纫线穿过几层面料形成连结线迹。

工业缝纫

虽然看上去这是一个简单的过程，但实际上工业缝纫是一个相当复杂的过程，包含了许多准备工作和为完美工艺而做的数学计算。优质的缝纫还取决于对纸样绘制的正确操作。(例如)撕拉平整布料上的洞，布会抽缩并形成复杂的立体形状，这就要求高水平的技术和经验来恢复面料的平整。对齐面料上的印花或织造图案也使设计过程复杂化。一旦一个设计师，根据他掌握的技术知识，编制了初步的工艺单和排料，随后面料就根据这些模型裁剪，并使用手工或机器缝制起来。

当整理面料并缝制时，面料必须保持挺括，没有褶皱。整个过程中，缝纫质量对布料不断变化的张力非常敏感。这些不良的变化影响产品的质量。因此,整个过程中严格控制非常必要。缝纫的工作重点就是控制面料,牵引面料沿着拼缝线在缝纫机上缝制。同时还需要保持最佳张力,才能确保高质量缝纫效果。

缝制前准备工作

在开始缝纫前,需进行下列准备工作:分开整理多层裁剪的裁片;将对应的裁片放置在工作台上;按正确的方向放置面料;调整面料的松紧。

在裁片分离过程中,送布装置持续不断地将成堆的裁片运往一个卸载位置,裁片堆上最上面的那层持续地被单独取出加工。在此过程中,裁片的一边被放置在夹具的夹片之间,旁边的开关调节夹片之间的距离。当这个距离控制在只允许一层裁片时,这层裁片即被传送到一个运送器的接收端进行下一步操作。根据气体动力学原理,最上面一层的裁片被吸附起来,从而与其他裁片分开。

当裁片放置在工作台上时,缝纫前还需完成的工作有:检查裁片形状、选择缝纫线迹,设计缝制工序和确定缝纫线。

检查裁片形状

根据面料的性能来确定缝纫张力。因此,需要根据裁片的物理性能来确定裁片的种类,例如是针织物还是梭织物等。面料裁片的不同形状:凸面的还是不凸面的,直线边形还是弧线边形,也需要考虑,每一种情况都需要不同的操作规范。总之,自动化缝纫设备的缝制方式要求面料(首先)分成若干个种类,并且必须遵守预先设计好的缝制顺序,才能得到理想的线迹。

缝纫线迹的选择

有两种基本的缝纫工艺:一种是连接两块衣片,另一种是装饰衣片。有时候,两种工艺不得不并用在服装的一些部分。例如,一个口袋必须有三边和服装缝合,同时还要有装饰线迹。在什么位置,必须使用什么种类的线迹——这些信息都通过计算机辅助设计系统数字化储存在自动设备中,用于指导缝制。

缝制工序的设计

缝制前就要确定缝制顺序。哪些衣片先被缝制,线迹的顺序是什么等。然而,有些工艺是必须在某些工序之前或之后的,例如上面的例子,装饰线必须先缝,其次是连接线。刺绣图案同样遵循这样的顺序,但有时候,某些服装类型,例如帽子、装饰线或刺绣就是在帽子完成后,用刺绣机加工的。

确定接缝线

缝制就是在面料内部距离布边几毫米,沿着接缝线缝纫。对于直线,接缝线就是裁片的外轮廓线往里面平移了,平移线的交点构成了接缝线的顶点。因此,接缝线与裁片的外轮廓线平行,两者之间的距离必须被确定,因为对于服装的不同部分,距离是不同的。裤脚的这个距离就比衬衫袖子的大。缝份就是这个布边和接缝线之间的部分。通常距离布边1.5厘米。用于标准家庭女裙缝制的缝份一般为2.5厘米或以上。工业成衣的缝份有所变化,但通常是0.6厘米。

缝制裁片

缝制过程主要包括三个步骤:引导面料穿过机针;缝合布边;围绕机针旋转。引导面料沿着接缝线前行,并与缝纫机保持速度一致。错误的方向不是被手工监控的就是被自动监控的，如果方向错误地输送进机器控制器，机器也会修正面料的方向。当接缝线的一条边被缝制时，面料会围绕机针旋转，直到下一条边也被连续缝制。缝制流程依此重复直到所有需要缝制的接缝线的布边均缝合好为止。

第二十一课　工艺单

工艺单是产品或服务需要严格遵守的标准。服装工艺单规定了原材料的要求、服装的制作过程，目的是为了达到公司设定的质量标准。为了使规格表易于理解，它们使用数值来表达。最大或最小的可接受数值或一个可接受的数值范围称为公差，允许与标准尺寸有一定的偏差。例如，就面料性能而言，理望的性能必须满足一个可测量的最大值，如 2% 的缩率。对于尺寸，通常使用一个公差。例如，男式衬衫的袖缝长为 32 英寸，允许上下公差 1/2 英寸。

一个服装制造商必须小心使用公差。例如，如果一件服装的所有尺寸均处于上公差范围，那么整个服装就可能太大。为了预防这种情况，制造商可以警告供应商，虽然服装的个别尺寸落在公差范围之内，如果整件服装的外形尺寸不在范围之内，服装仍然会被拒绝接受。这个警告必须通过某种可以测量的量值或比例来表明。

就标准性而言，工艺单的编写必须准确、前后一致，供应商和采购商都能完全准确理解。为了达到这个精确度，只有受过专门训练的人才会被雇用编写工艺规格表。工艺单的编写人必须对产品的所有尺寸有充分的了解，并且细心、严格、条理清晰、准确。面料规格表编写人要精通纤维组成、精纺工艺、针织工艺、梭织工艺、色彩学以及化学和后整理。

工艺规格表是交流服装工艺规格的文件，公司内外均可使用。工艺规格表的内容组成来自各种原始资料。

工艺规格表首先用于产品生产阶段。当样品被制作出来时，设计师需要就样衣的外观与纸样师和样衣工准确地沟通。

规格表还将主要的信息传递回制造环节。通常规格表是一个文件包的一部分，文件包还包括产品样板、一套纸样、可能还有生产的排料图。这个规格表应该包括下列内容：

- 款式标识；
- 效果图或照片；
- 各档尺码，每码的所有尺寸以及公差；
- 色彩设计；
- 面辅料、附件的使用率；
- 结构的细节包括缝线和缝份、缝法、每英寸针量、部件的位置，例如标签，口袋等；
- 参考工序表；
- 水洗标签所包括的信息。

第二十二课　质量评估

质量可以根据成品的品质来判定，也可以根据加工中的工艺特点来评估。

理论上，质量评估从纸样的选择即开始，一直到最后的试样和熨烫。一个纸样的选择应该适合个体对象的特征和体型，假如这样的纸样难以在商业纸样中找到，可以组合各个纸样部位，或根据对象单独设计。

时装领域的许多评判不在于观察服装的内部结构，因为他们笃信任何服装外观上的不足均能在内部结构上找到原因。例如，服装上的一条缉线，由于内部缝制不到位，模特试穿时，会在服装表面显露出来。

在评估服装的结构方面，服装工艺制作的方法很重要。它必须适合面料纤维的成分和种类、服装的款式以及服装的用途。

服装结构综合评价得分表

综合分数		分数	中等	良好	优秀	评价
面辅料选择的匹配度	与设计对象匹配度					
	与纸样的匹配度					
	与工艺的匹配度					
	缝线 — 适合的种类 — 适合的色彩					
	与时尚面料的匹配					
	衬布和衬里 —重量 —位置					
	里布 —色彩 —重量 —与面料的匹配度					
工艺的整洁度与规范度	机器线迹 —适合的针距 —均匀的张力					
	缉线处理 —与面料匹配 —与面料的位置相称					
	手缝工艺 —合适的线缉 —长度均匀 —合适的长度					
	熨烫工艺 —无极光 —无烫痕 —无褶皱 —线缉熨烫到位					
	线缉准确度 直线线缉 —线缉到位 —线缉顺直					

续表

	弧线线缉 —线缉到位 —线缉圆顺					
	裁片对应的边线长度与宽度相当 —领尖 —大身缝线的长度 —裙片缝线的长度 —对应的省道 —肩线 —钮洞 —装饰细节					
	服装各部位匹配度 —对称的省道 —格子 —条纹 —图案 —花卉					
综合外观	省道 —省道形状 —安全缝缉 —熨烫到位					
	贴边 —缝迹使用刀眼标记 —缝份宽窄分级 —暗缝线迹 —衬布加固					
	腰线处理 —定位准确 —固定位置针迹合理 —缝线熨烫方向					
	碎褶 —褶量均匀分布 —不出现塔克或褶裥					
	褶裥 —褶量均匀分布 —宽度均匀					
	钮洞 —合适的位置 —合适的尺寸 —边缘均匀,尺寸标准					
	口袋 —位置 —折边均匀 —正确的尺寸 —口袋折边适当 —贴袋位置 —使用的结构类型					

续表

	拉链 —表面线迹宽窄均匀 —里襟隐蔽 —拉链嵌缝中间闭合 —拉链使用长度 —安装类型 —与面料颜色匹配					
	袖片 —装袖吃势圆顺 —连裁袖加固					
	领口和袖克夫 —缝迹使用刀眼标记 —缝份宽窄分级 —暗缝线迹或领下片斜裁					
	下摆 —宽度均匀 —合适的宽度 —面料后整理 —正确的服装款式 —使用接缝贴边,色彩与面料匹配 —服装正面的外观效果					
	腰带 —合适的宽度 —整洁的处理					
	紧固件 —正确的位置 —合适的尺寸 —配套的色彩 —安全紧固 —包布按扣 —钮扣柄长适当					
	补充 —手工拱针 —同料饰边					

第二十三课　标签和吊牌

现在每件服装必须有一个或多个标签,给消费者详细的信息。这些信息是法定的,必须执行。这些标签必须放在服装上易于发现的地方。通常,放在服装的后中心:衬衫、毛衣或短衫的领部,裤子和裙子的腰部。有时你也会在茄克的前里缝底部或内衣的内侧缝看到。

这些商标可以粘贴、缝制、印制或拓印在服装上。他们甚至被放在服装的外部,一旦被放置后,就一直在那里,直到你——购买者决定取下它为止。

但是，大多数标签永久附着在服装上。如果服装一直被包装在袋中，直到售出才能取出，这件服装的纤维成分标签就只能粘贴在包装袋上。

根据纺织纤维产品信息条例有5种必须出现在服装标签上的强制信息：

1.服装上任何重量占5%或以上的纤维成分必须标出；

2.按重量列出纤维含量百分比；

必须按降序列出纤维百分比。首先是主要成分的纤维，最后是成分不足5%的“其他纤维”。如果服装只使用了一种纤维，标签必须列出，它可以标成“100%棉”或“纯棉”。

3.厂商识别

这个内容将标明谁为产品负责。制造商或商场、登记注册号或产品的商标都写在标签上。商标是一个符号、设计、文字或字母，是由制造商或零售商用来与同类竞争者相区别的。商标被注册，即受法律保护，没有人可以使用他人的商标。

4.出产地

标签必须标明服装的产地，如“印度制造”或“香港制造”。

5.使用须知

有关服装的护理信息，由美国联邦贸易委员会颁布的水洗标签规则规范。

吊牌是吊挂式文字描述的服装标签。他们可以被一根线、一条塑料片或一根别针固定在产品上。在你穿用这些服装之前，拆除这些标签。吊牌可能会重复服装标签的一些信息。但是，在吊牌上的信息是随意的，自由而不受法律约束。

制造商的吊牌可能包括以下信息：

- 制造商的商标或品牌名称，这可以采取标识或符号的形式，表现产品或制造商；
- 商标、品牌名称或纤维的标识；
- 面料组织结构的信息。此信息可能会标明面料是弹性织物或是针织物；
- 有关服装或其面料保修的信息。

第二十四课　时装市场的组成

除了工艺上的原因之外，时装之所以能被整个大众市场接受，是因为根据服装的功能，存在几个档次：

- 高级订制；
- 设计师品牌；
- 街头大众/服饰市场。

高级订制

它们是举世闻名的高级作坊，由国际上认可的著名设计师主持。高级订制每年至少两次展示作品秀，并高价出售度身定制的单品。对于这些设计师而言，时装秀实质上就是推广他们的服装、香水及配件的宣传活动。

近些年来，高级订制师和知名服装设计师均开始出现细分品牌和资本化品牌的迹象，有些品牌已经开始分散制作加工来降低成本。

一些制造商筹备和推广设计师发布会，这使得高级订制或设计师品牌服装能以成衣的形式在巨大的低价市场销售。1922年法国政府颁布的鼓励新设计师进入高级订制业的声明，就反映出高级订制正在衰退。

年轻的客户因为喜好逐渐转向同时代具有共同方向和生活方式的设计新秀们，并逐渐远离了奢侈的产品。

设计师品牌

设计师品牌通过高级女装成衣来表现。设计师进入成衣市场意味着他们能为更多的消费者提供风格化的设计和高品质产品。这类服装可以在设计师门店、专卖店和商场专柜看到。

设计没有了限制，但是生产的数量仍然被控制。高级时装成衣设计师和高级订制设计师有着共同的目标，就是设计优美并吸引人的时尚款式。高级成衣设计通常创新较少，并模仿高级订制的设计和风格，他们可能会根据一年前卖得很好的某款外套设计类似的产品。

大众市场或街头服饰

这是大多数人购衣的市场。新潮时装很快在大街的商店里看到。这是一个容纳众多变化的市场。

市场的三等级观的划分也许过于简单了，因为在这些市场中还可以细分成许多档次和价位。

许多消费者并不坚持只买一个档次的服装，更多的热衷高级订购的富人们，转向设计师品牌来打点他们的日常装。经常购买设计师品牌的女性也可能为某些特殊的场合，花费大量金钱购入一套高级订制的裙装。那些通常只购买成衣的人，会偶尔在折扣店买打折的设计师品牌。

二手服装在高级社区中心的慈善商店出现的频率和在新品店差不多。由于数量上的激增，二手服装可以通过多种渠道进行销售。

第二十五课　服装贸易

购买计划

贸易的第一步是拟定购买计划。因为业务员负责提供一个反映了消费者的需求、尺寸和销售季节的清单，因此设计一套预算、一个购买计划是非常必要的。预测市场需求和尺寸称为时尚预测。购买计划通常以六个月计，因此被称为六月购买计划。购买计划为每一个类别列出预期的销售额以及去年的销售额。一个类别就是一种产品，例如针织上装或牛仔裤。此外，购买计划还包括许购定额。许购定额是指采购商在一定时间内可以用来购买产品的总额度。为了一切有实用价值，许购定额被采购员用于购买货物的金额。许购定额是根据预期的销售额、理想的库存水平和已有的订单而制定的。许购定额为采购商提供了业务标准，避免购买过量和购买不足。

采购和订单

一旦知道了他们必须要付出的，采购商随即与设计师合作，根据公司的类型，或参加交易会为新时装的生产联系订单。在这个过程中，一个精确的计划非常必要，会避免采购商高买低卖。大多数采购商一年会至少进行两次这样的联系过程，一次是秋/冬季、一次是春/夏季。采购商可以在两种地方进行交易，一是他们可以参加展销会，一般位于时装中心；时尚中心是一座云集众多展销会或陈列室的城市，还有一些专业展览会，例如婚礼服展。陈列室是采购商可以去订购商品的另一个地方。陈列室通常由设计师或服装品牌的销售代表运行。

采购商也常被带着各种系列设计和产品册的生产厂家的代理造访。理想的采购商，需要对流行趋势有深刻的研究，了解企业发展的方向，了解商场产品价格的最佳范围。这样，在买卖和订单交易的过程中，采购商会牢记这些信息。

接受订单、为客户提供信息

根据公司的类型，采购商或商品规划主管将负责确认货品到店。最终在店内的货品要符合当地的气候以及店铺的大小。此外，采购商还需与新产品和视觉陈列团队以及销售部门进行沟通。这些会在新产品会上正式或非正式地沟通解决。

库存监督和销售

根据公司的规定，监督库存水平可能是商品规划主管的工作。商店常连续地使用复杂的扫描设备和资产管理软件。这些软件可以防止盗窃，但它还提供有关库存水平的采购商的实时信息。任何规模公司的采购商，了解商场里的货品销售情况都至关重要。对于那些销售良好的产品，采购商会考虑翻单，而那些滞销的产品，采购商则会计划分销或打折销售。许多零售商拥有复杂的计算机系统来监测库存情况。

回购、广告的协商

贸易商最后的责任是和供应商谈判，这需要重点说明。虽然，谈判贯穿整个时尚购买过程，如买家在购买和订购阶段与供应商进行价格的协商，但是，还有很多其他需要谈判解决的问题。首先，采购商会协同销售部门就公司的广告与卖家进行协商，目的是削减广告成本，并保证广告能体现制造商服装的特点。另一类谈判可能是就回购滞销产品进行的协商，比如，采购商选中的供应商承诺的橙绿色衬衫，而实际提供的货物色彩过于强烈，导致滞销问题。

此外，贸易商还将就供应商免费或低价提供卖点展示进行谈判。

第二十六课　标准

服装公司已经意识到整个商业操作中质量管理的重要性，并设立相应的生产和销售的预期值。标准是对于在产品研发和制作过程中广泛交流的定量或定性的数值，建立的可接受的程度的描述。这些说明是对服装多种元素的预期值的描述：如尺寸、材料、款式部件、外观和性能等，这些元素应用在原材料、设计、生产和包装中。所有影响服装质量的员工都必须清楚理解标准的术语含义，这一点至关重要。有时候，需要使用插图、照片、图表、定义和举例来辅助说明。

质量要求应通过书面形式来交流，例如一种分发给公司员工和外部供应商及承建商的质量保证手册。质量保证手册应该标注出准确严密的产品性能，包括性能的最低程度和理想程度。

标准化是为开发符合标准要求的产品的公司建立规章的过程。标准化需要供应商、经营商和消费者达成一致。当整个行业接受同一个标准的制约，这个标准被称为是行业标准；当某个标准被设定，但不是所有的公司都遵循这个标准，此标准被称为是自愿执行标准。在引发公共安全问题时，例如考虑到婴儿睡衣的可燃性、抽带和小部件的使用等，政府成为标准执行中的首要角色。如果标准是法定的，则被称为是强制性标准。

在一个标准被所有行业人员接受之前，它们被检验、评估，可能多次修改。由于标准为整个供应链提供了一个评估和比较同类产品的方法，因此标准是鼓励竞争的。例如，前中口袋的缝制：高级衬衫采用每英寸 22 针的缝制工艺，而每英寸 10 到 12 针则用于缝制较廉价的衬衫。业务员必须认真对待标准的作用以及在研发制造中的产品标准化。

第二十七课　产品检验

许多时装零售商为生产厂家提供质量手册，确保双方都非常清楚产品的指定标准。

零售商的服装技术员可能负责编写、审阅或更新公司的质量手册，考虑法律或目标市场的变化。质量手册涉及法规、健康和安全须知以及环境保护等因素。随着涉外贸易的增加，质量手册还需适当修改，体现不同国家的操作要求，以确保沟通通畅。在检验中，服装技术员需警惕的劣质产品的特征有：

- 起皱或断断续续的线迹；
- 跳线；
- 紧固件和配件的安全性；
- 口袋或紧固件的位置；
- 松线。

供应商的服装工艺师在制作中负责检查成衣的质量标准，并按照零售商的要求寄出样衣。面料检测是时装零售商质量控制的内容之一。如果公司设有面料工艺员这个职位的话，检查供应商提供的面料的检测标准，是零售商的服装工艺员或面料工艺员的工作之一。

零售商的技术员有时会去厂家验货。这个验货被称为跟单，包括对正在生产和准备销售的产品质量标准的评估。跟单有助于降低被拒次品的回单率，节约时间和成本。质量检查包括测量服装的各种尺寸，确保它们符合要求的规格表，评估生产质量。当面料是一种弹性材料时，根据规格表上相应服装部位的“容差”规定，允许 0.5 厘米到 1 厘米的微小的误差。

服装工艺员要检查所有自己负责的服装是不可能的，尤其是那些海外加工的产品。这导致了供应商自行认证的增加，采用一个专门的检查系统，使供应商有更多的责任感，使产品能够符合零售商的质量标准。

独立的检测实验室也能签订合同，代表零售商去执行跟单检查。

成衣被运往零售配货中心，通常位于一个中心地带或分散几处。零售服装技术员常常去配货中心，在产品发往各个零售点之前，检查大货质量。零售商常常聘用销售中心资深的质检员，他们与总部技术部门有一定的联系，质检员随机检查一小部分服装。如果出现质量问题，这些款式的服装就要求在仓库里全部验货，如果质量一直很低，交货可以被拒。

第二十八课　纸样绘制系统

计算机辅助设计系统是工程师和设计师众多的使用工具之一，根据使用者的专业和软件种类的不同被广泛使用。计算机辅助设计软件有好几种，每一种都要求操作者分别考虑他们的使用方法，操作者必须按照不同的方式为每个内容设计虚拟组件。

纸样生成软件

纸样生成软件是一个问答模式的纸样设计软件，与使用绘图工具绘制一个新的衬衫领不同，纸样设计师通过输入文字和尺寸要求在屏幕上设计款式的纸样。

许多不同服装的部件被存储在系统中，设计师可以连接和调整不同的组件以形成效果图。

例如，调出大身的前后片 X，与袖片 Y 和领片 Z 拼接。如果肩线需要降低，操作者可以指令计算机降低 0.5 英寸,计算机不会只缩短大身的尺寸,它会自动调整相联的袖片和领片。这里,已存储的所有衣片均被检测了匹配度。这样,样板绘制过程就被缩短或可以省略。新款式的纸样完成后,计算机可以生成一整套全码的样板。

放码系统

在一个样品尺寸纸样被输入之后,需要进行推档放码。纸样上某些点被认为是"放码点"或排料位置,在这个位置,样板按照一定规则进行放缩。在每一个放码点,操作者将放码规则输入计算机,这些放码规则指令计算机在 X、Y 轴上变动进行尺寸放缩。这个系统将自动产生预定的各档尺码。随后,纸样的各个部分从最小到最大码,一个套一个,或称为嵌套叠放,这样,打样师一眼就能发现放码中出现的任何错误。

排料系统

排料就是针对某款服装的所有裁片的最优化方式排列,目的是争取面料最大利用率。在排料系统中,操作者在屏幕上移动所有尺码的裁片小样,以获得最小的损耗率。为避免混淆,每个样板以不同的颜色标注尺寸,系统会通过指令使操作者了解排料损耗率。有些系统在计算机纸样排列系统中引入算法,这样操作者参予排料的工作被减少。条纹布或格子布以及倒顺毛面料在排料时需要考虑面料的幅宽。大多数系统都有自动对格对条的功能,如果操作者想要从某个角度旋转纸样,使得排料更加紧密,这也容易实现。

绘图功能与排料系统相联,允许排料图在大型的滚筒纸上以不同的尺寸打印。这张纸将被放在铺好的面料上进行裁剪。个别的全尺寸或者半尺寸的裁片和套裁裁片一样使用打印机打印出来。

裁剪操作系统

裁剪操作系统与打板系统相联,裁剪操作系统需要计算裁剪的面料层数以适应特定款式的色彩和尺寸的需要。裁剪系统可以不需要打印出排料图,而是将排料系统生成的纸样形状的信息直接传送给裁剪机。裁剪机自动裁剪,无需人力。裁剪路径也可由电脑计算得出,以获得最有效和准确的结果。

第二十九课　工艺单和成本软件系统

这个系统涵盖了款式的所有信息,包括平面效果图、尺寸规格、裁剪要求和尺码表,这些文件统称为规格表或工艺单。这套软件跟踪所有有关的设计、生产的信息,必要的话,甚至包括产品折叠方法和船运说明。当公司有一系列加工点时,工艺单和成本软件系统有助于确保准确性和连贯性。这特别有助于那些需要海外加工的公司,甚至是必配的。

许多公司都经历有关文化和语言障碍的沟通问题,CAD 可视系统使误解降到最低。

规格表管理是计算机辅助设计系统极为重要的一块内容,因为它涵盖所有生产、成本和出货的文件。当设计师制作一件服装时,所有相关的成本,甚至缝线的耗用量,都被立刻确定下来。设计或缝制顺序中任何微小的变化,都能被立即发现。服装制造商的工艺单系统一般都持续网络在线,这样,公司任何人均可登陆信息,无论他们身处何地。

规格表和成本核算系统也可以链接到库存控制系统。他们从排料系统和试生产系统中添加信息,并跟踪从购买面辅料开始一直到最终产品的所有生产数据。规格表和成本核算系统将大大小小的工厂和零售商联接起来,扩展了信息网络。

第三十课　未来展望

未来产品开发中，数据系统将扮演一个重要的角色。随着计算机发展速度的加快，三维设计系统和模拟图像技术将大多数产品开发过程转变成高速数据化应用处理过程。

使用计算机虚拟技术制作模拟样品，基于互联网技术的先进视听通讯系统，将支持来自世界各地的产品开发团队对样品进行评估。

越来越多的计算机辅助应用程序中添加了二维和三维的立体仿真工具，不需要制作出样品，设计师们就可以通过创建计算机合成图像来评估造型设计的构思。纺织计算机辅助设计程序支持生成虚拟梭织、针织面料，使得模拟样品更加真实。

计算机辅助设计/计算机辅助制造系统的重点是针对开发中所有环节，提供成套的程序作为研发方案，它还包括第二代系统：

- 二维和三维的创意绘制软件，模型软件，立裁软件，模拟真实款式的虚拟样品，并进行评价。
- 纸样绘制工具，可以将三维模拟产品转化成二维平面纸样，并提供匹配的面料机械性能和放松量。
- 纺织计算机辅助设计系统能生成模拟样品上的各种梭织、针织和印花图案，通过与面料商、印花商们的网络共享，缩短面料采购周期。
- 织物数字印花机支持服装设计师使用印花面料制作真实样品。
- 面向供应商的 B2B 电子商务支持贸易商和设计师在线浏览面料、辅料的目录，并使用这些原材料虚拟出样品，实现电子下单。
- 数据管理系统将收集所有工艺细节、图纸以及效果图，并为新款的开发保留完整的数据，同时在整个产品的开发过程中，使用互联网传播信息。

当基于互联网技术的低成本视频会议快速变成一种现实时，产品开发团队之间的合作更像是数据的交流，而不是人与人之间的互动。洛杉矶的贸易商可以联系纽约的设计师，一起浏览和讨论泰国曼谷某合作工厂内试衣模特身上的一件样衣。

通过高分辨率的视频、数据压缩以及高速的互联网传输，实时电信会议使产品开发团队能够放大观察服装上极小的结构细节，或模特着装后的服装的光滑感。语言翻译软件将提供各种语言的同步翻译。这将消除引发大量全球服装采购问题、阻碍产品合作开发的语言障碍。

单词表

A

a whole host of 大量,众多
abrasion resistance 抗磨损性
absorbency [əbˈsɔːbənsi] *n*. 吸收性,吸收率,吸收能力
accessory [ækˈsesəri] *n*. 服饰品;配件,备件;辅助设备
accommodate [əˈkɔmədeit] *v*. 使适应,向……提供
accomplish [əˈkɔmpliʃ] *v*. 完成,贯彻,实行
accredited [əˈkreditid] *adj*. 可接受的;可信任的,被正式认可的
acetate [ˈæsiˌteit] *n*. 醋酸
acrylic [əˈkrilik] *adj*. 丙烯酸的
active [ˈæktiv] *adj*. 活跃的,积极的,精力充沛的
adhesive [ədˈhiːsiv] *n*. 粘合剂
adjunct [ˈædʒʌŋkt] *n*. 附属物,附件,助手
adjustment [əˈdʒʌstmənt] *n*. 调整,调正,整理;修正
advertising agency 广告公司,广告社
aerodynamic [ˌεərəudaiˈnæmik] *adj*. 空气动力学的,气体动力学的
aesthetic [iːsˈθetik] *adj*. 美学的,审美的,有美感的
affiliation [əˌfiliˈeiʃən] *n*. 联系,归属,关系
affinity [əˈfinitiː] *n*. 亲和力
affix [əˈfiks] *v*. 粘贴
affluent [ˈæfluənt] *n*. 富人
agent [ˈeidʒənt] *n*. 代理人,代理商
agreed-upon 达成协议的
air vents 通风口透气孔
airy [ˈeəriː] *adj*. 透气的
akin [əˈkin] *adj*. 同类的,同族的,同源的
alert [əˈləːt] *adj*. 警惕的;警觉的,留心的
algorithm [ˈælgəriðəm] *n*. 算法;规则系统;编码
A-line A 字形线条,A 型造型
allocate to 分派,配给
alter [ˈɔːltə] *v*. 修改,改动;改变
ambiguous [æmˈbigjuəs] *adj*. 不明确的,模棱两可的,模糊的
ample [ˈæmpl] *adj*. 容量大的
an array of 一批,一系列
analyze [ˈænəlaiz] *v*. 分析,细察,分解
ankle [ˈæŋkl] *n*. 踝骨,脚脖
announce [əˈnauns] *v*. 发布,宣布,预告,通知
announcement [əˈnaunsmənt] *n*. 公告,发表,告知
anorexia [ˌænəˈreksiːə] *n*. 厌食症
antagonistic [ænˌtægəˈnistik] *adj*. 对抗的,不相容的
anthropologist [ænθrəˈpɔlədʒist] *n*. 人类学家
anticipate [ænˈtisipeit] *v*. 预期,希望,预料
apparatus [ˌæpəˈreitəs] *n*. 仪器,装置,器皿,用具
appealing [əˈpiːliŋ] *n*. 吸引力
appearance [əˈpiərəns] *n*. 外观,外貌
applicability [ˌæplikəˈbiləti] *n*. 适用性,适应性
applicable [ˈæplikəbl] *adj*. 合适的,可应用的
applique lace 贴饰花边
appraise [əˈpreiz] *v*. 评价,估价
appropriate [əˈprəupriət] *adj*. 适当的,适用的

approval [ə'pruːvəl] *n*. 批准,认可,确认
archaeologist [aːkiə'lɔdʒist] *n*. 考古学家
arise [ə'raiz] *v*. 出现,发生
arm [aːm] *n*. 袖子;手臂,上肢
armhole line (AL)袖窿弧线
armhole ['aːmhəul] *n*. (AH)袖窿
arrangement [ə'reindʒmənt] *n*. 整顿,布置,排列,商定,协议
art critic 艺术批评家
artifact ['aːtifækt] *n*. 文化遗物,遗迹
artificial silk 人造丝
artistic achievement 艺术成就
artwork ['aːtwəːk] *n*. 插图,艺术作品
ascot tie 阿司阔领带,领巾式领带
assembly line 装配线,生产流水线
assessment [ə'sesmənt] *n*. 评估
asset ['æset] *n*. 有用的资源,宝贵的人[物]
assistant designer 助理设计师
assorted [ə'sɔːtid] *adj*. 分类的,配色的
assumption [ə'sʌmpʃən] *n*. 假定;设想
astonish [əs'tɔniʃ] *v*. 使……吃惊
asymmetrical [ei'simitrikəl] *adj*. 不对称的
at random 随机地
athletic [æθ'letik] *adj*. 运动的,身体健壮的,活跃的
attachment [ə'tætʃmənt] *n*. 附件
attendee [ˌæten'diː] *n*. 与会者
attribute [ə'tribjuːt] *v*. 认为……属于
attune [ə'tjuːn] *v*. 调(音),调谐使调和,使协调
audiovisual [ˌɔːdiːəu'viʒuːəl] *adj*. 视听的
automate ['ɔːtəmeit] *v*. 自动作业,使自动化
automatic [ˌɔːtə'mætik] *adj*. 自动的
available [ə'veiləbl] *adj*. 可利用的,可获得的,有效的

B

B2B Business to Business 企业到企业的电子商务模式
back and forth 来回地
back bodice 后衣身
back pitch 后袖窿吻合点
back [bæk] *n*. (织物)背面
backing ['bækiŋ] *n*. 衬里,底布,背衬
baggy ['bægi] *adj*. 膨胀的,凸出的
balance mark 对位记号
balance ['bæləns] *n*. 天平,秤,平衡
ball gown 晚会礼服,正规礼服
band knife 带刀
barrier ['bæriə] *n*. 障碍
based upon 基于……
basic block 原型
basic seam 基本线,基本缝纫结构线
baste [beist] *v*. 用长针脚粗缝
basting cotton 粗缝棉线,绗缝棉线
batch [bætʃ] *n*. 批,群,组;成批生产
batik printing 蜡染
be rooted in 深植于
beadwork ['biːdwəːk] *n*. 珠绣
bean-shaped 豆圆形的
beeswax ['biːzwæks] *n*. 蜂蜡,蜂蜡色
bell [bel] *n*. 喇叭口;降落伞衣;钟形物
belt [belt] *n*. 带,皮带,腰带,肩带
bent-handle 弯把
bewildering [bi'wildəriŋ] *adj*. 令人困惑的,令人不知所措的
bias ['baiəs] *n*. 斜纹路;斜条; *v*. 斜裁
bicep-upper arm girth 上臂围
bill of materials 材料单
Bk. back 的缩写
bleach [bliːtʃ] *v*. 漂白,褪色
bleed [bliːd] *v*. 渗出;(印染等)渗色,渗开
blend [blend] *v*. 混合
blind hemmer 暗卷边机
blind-stitching 暗针,暗缝线迹
blister ['blistə] *v*. 起泡
block printing 木板印染法

block [blɔk] *n.* 原型,裁剪样板
blouse [blauz] *n.* 衫;上衣;罩衫;宽大短大衣
board [bɔːd] *n.* 熨衣垫板
bobbin ['bɔbin] *n.* 筒管,筒子
bodice ['bɔdis] *n.* 上衣片,大身,女装紧身上衣,紧身胸衣
bodkin ['bɔdkin] *n.* 穿带用的钝针,粗长针;锥子
body measurement software 人体测量软件
body scanning software 人体扫描软件
body stocking 女子贴身连衣裤,紧身连裤袜
bold [bəuld] *adj.* 鲜明的;清晰的;醒目的;大胆的,勇敢
bolo tie 保罗领带,饰扣式领带
bonding ['bɔndiŋ] *n.* 连接,结合,加固,粘合
border ['bɔːdə] *n.* 边纹,镶边,滚边
bottom ['bɔtəm] *n.* (织物)底色,下方;(衣服)下摆,下装
bounce back 迅速恢复活力
boundary ['baundəri] *n.* 界线;边界;境界;范围
bow tie 蝴蝶结
bracelet ['breislit] *n.* 腕饰物,手镯,臂镯
braid [breid] *n.* 辫子,发辫,饰带,编织物,滚带
breathability [breθə'biləti] *n.* 透气性
breed [briːd] *v.* 繁殖,饲养,产生
bridal wear 婚礼服
bride [braid] *n.* 新娘
brooch [bruːtʃ] *n.* 胸针,领针,扣花,胸饰
brush [brʌʃ] *v.* 擦,刷亮
buckle ['bʌkl] *n.* 扣子,带扣
budget ['bʌdʒit] *v.* 预算,安排
bulk fabric 大货面料(大批量生产使用的面料)
bulk [bʌlk] *n.* 膨松度
bulkiness ['bʌlkinis] *n.* 膨松性,膨松度
bundle ['bʌndl] *n.* 捆,束,包 *v.* 捆,扎
bust line (BL)胸围线
bust point (BP)胸高点,乳点
bust [bʌst] *v.* 胸围,胸部
button ['bʌtən] *n.* 扣子,钮扣,按钮
button-down shirt (前开襟钮扣的)传统衬衫
button-hole twist 锁钮孔线
buttonhole ['bʌtnhəul] *n.* 扣眼
buyback [bai'bæk] *v.* 回购,产品返销
by-product 副产品

C

calender ['kælində] *n.* 轧光整理
calico ['kælikəu] *n.* 平布,白布,印花布,棉布
camera ['kæmərə] *n.* 照相机
camouflage ['kæməˌflaːʒ] *n.* 伪装,掩饰,保护色,迷彩
campaign [kæm'pein] *n.* 竞选运动
canvas ['kænvəs] *n.* 帆布
capitalize [kə'pitəlaiz] *v.* 使资本化,估计……的价值
capture ['kæptʃə] *v.* 捕获,捕捉
car shows 车展
cardigan sweater 卡蒂冈毛衫,开襟式毛衫
care requirement 使用须知
carve [kaːv] *v.* 切割,雕刻
casein ['keisiːin] *n.* 酪朊,酪蛋白;酪素
casing ['keisiŋ] *n.* 抽带管;装嵌条
casual-wear 便服,轻便装
catalog ['kætəlɔg] *n.* 目录
categorization [ˌkætigəri'zeʃən] *n.* 归类,分类
category ['kætigəri] *n.* 种类,部属;类目

cater for　投合，迎合
catwalk [ˈkætwɔːk] *n*. (舞台等)天桥
celebrity [siˈlebriti] *n*. 名人，名流
cellular device　手机
cellulosic fiber　纤维素纤维
centerfold [ˈsentəɪfəʊld] *n*. 中间折叠
central back line　(BCL)后中心线
certificate [səˈtifikit] *n*. 证(明)书，凭证
certification [ˌsəːtifiˈkeiʃən] *n*. 证明，保证，鉴定证明书
characteristic [ˌkæriktəˈristik] *adj*. 表现特点的，特有的，表示特性的
charity [ˈtʃæriti] *n*. 慈善机构
chartreuse [ʃɑːˈtrəːz] *adj*. 鲜嫩的黄绿色
chemise [ʃiˈmiːz] *n*. 直统连衣裙，女式无袖衬衣
chequered [ˈtʃekəd] *adj*. 格子花纹的，方格图案的
chest width　胸宽
chest [tʃest] *n*. 前胸
chevron [ˈʃevrən] *n*. 山形，人字形；波浪形；V形
children-wear　童装
chlorine [ˈklɔːriːn] *n*. 氯
chop [tʃɔp] *n*. 商标
chroma [ˈkrəumə] *n*. 纯度；色品；色度
circular [ˈsəːkjulə] *adj*. 圆形的，循环的
clammy [ˈklæmiː] *adj*. 滑腻的，粘糊糊的，冷淡的
classic [ˈklæsik] *n*. 不受时尚影响的，经典的(服装)，传统(服装)
clay [klei] *n*. 泥土，黏土
clergy [ˈkləːdʒi] *n*. 牧师，僧侣，神职人员
clientele [ˌklaiənˈtel] *n*. 客户
clinging [ˈkliŋiŋ] *adj*. 紧身的，贴身的
clip [klip] *n*. 别针，首饰别针　*v*. 剪去，修剪，剪短
clip-on tie　夹式领带，夹式领结
cloche hat　钟形女帽
cloque [kləuˈkei] *n*. 泡泡纱
closure [ˈkləʊʒə] *n*. 服装闭合件
cloth [klɔ(ː)θ] *n*. 织物，布，衣料
clothe [kləʊð] *v*. 给……穿衣，盖上，赋予
clothing [ˈkləuðiŋ] *n*. 服装，服饰，衣着，衣服，衣饰
cloth-lay　铺料
club-wear　俱乐部服装，娱乐服饰
clustered [ˈklʌstəd] *adj*. 成串的，成群的，成串的；聚集
coarse [kɔːs] *adj*. 粗的
coating [ˈkəʊtiŋ] *n*. 上胶，涂层，涂料
cocoon [kəˈkuːn] *n*. 茧，蚕茧
coded mark　编号，记号
coincide with　相符，与……一致
collaborate with　合作
collaboration [kəˌlæbəˈreiʃən] *n*. 合作
collar [ˈkɔlə] *n*. 领，衣领，上领
collection [kəˈlekʃən] *n*. 服饰系列，季节服装系列；时装展览，时装发布会
color variation　色差
color-block　色块
colorfastness [ˈkʌləfaːstnis] *n*. 染色坚牢度
comb [kəum] *n*. 梳　*v*. 梳理
combine [kəmˈbain] *v*. 使联合，合作
come off　发生，表现
comfort [ˈkʌmfət] *n*. 舒适，舒适性
commercial placement　商业纸样的处理方法
commercial [kəˈməːʃəl] *adj*. 商业的
compare to　与……相比
comparison [kəmˈpærisən] *n*. 比较，对照
compass [ˈkʌmpəs] *v*. 达成，完成
compatible with　协调的，相容的
competition [ˌkɔmpiˈtiʃən] *n*. 竞争，竞争者
competitiveness [kəmˈpetitivnis] *n*. 竞争
compile [kəmˈpail] *v*. 编制；搜集
comprise [kəmˈpraiz] *v*. 包含，由……组成

computer-generated 计算机生成的
computerize [kəm'pjuːtəˌraiz] *v*. 给……装备计算机
concentrate ['kɔnsəntreit] *v*. 集中;使……集中于一点
concisely [kən'saisli] *adv*. 简明地
conduct ['kɔndʌkt] *n*. 管理,处理
cone [kəun] *n*. 锥体,锥形
configuration [kənˌfigju'reiʃən] *n*. 结构;构造;布置方位;构型,排列
confine [kən'fain] *v*. 限制,局限
conflict ['kɔnflikt] *v*. 战斗,冲突,矛盾,抵触
conform to 符合,遵照
conformity [kən'fɔːmiti] *n*. 相似;遵从,顺从
confuse [kən'fjuːz] *adj*. 混乱的,混淆的
confusion [kən'fjuːʒən] *n*. 混乱,混淆
consensus [kən'sensəs] *n*. (意见等的)一致
considerable [kən'sidərəbl] *adj*. 重要的,不可忽视的
consideration [kənˌsidə'reiʃən] *n*. 考虑,思考
consistency [kən'sistənsi] *n*. 始终一贯;前后一致
consistently [kən'sistəntli] *adv*. 一贯地,一向,始终如一地
constantly ['kɔnstəntli] *adv*. 不变地,不断地,时常地
constrain [kən'strein] *v*. 抑制,约束
constrict [kən'strikt] *v*. 压缩,收缩
construct [kən'strʌkt] *v*. 构造,结构
construction line 结构线
consultant [kən'sʌltənt] *n*. 顾问
consumer [kən'sjuːmə] *n*. 消费者;用户
consumer-product company 消费产品公司
contemporary [kən'tempərəri] *adj*. 同时代的;当代的
contour ['kɔnˌtʊə] *n*. 轮廓,外形;轮廓线;略图
contractor [kən'træktə] *n*. 承建商,承包商
contribute to 捐献,贡献
contributor [kən'tribjuːtə] *n*. 贡献者
convenient [kən'viːnjənt] *adj*. 方便的,便利的
convex ['kɔnveks] *adj*. 凸的,凸面的
convey [kən'vei] *v*. 传达,运输,转让
conveyor [kən'veiər] *n*. 传播者,运送,运输设备
coordinate [kəu'ɔːdineit] *v*. 坐标
cop [kɔp] *n*. 纺锤状线团
cord [kɔːd] *n*. 线绳,滚条
cording foot 嵌线压脚,滚边压脚
cording ['kɔːdiŋ] *n*. 嵌线,包梗
corduroy ['kɔːdəˌrɔi] *n*. 灯芯绒
corporation [ˌkɔːpə'reiʃən] *n*. 公司,企业
cost sheets 成本核算表
cotton gin 轧棉机,轧花机
counterpart ['kauntəpaːt] *n*. 对手
countless ['kaʊntlis] *adj*. 无数的,数不尽的
country of origin 原产国,原产地,出产地
couturier [kutyr'je] *n*. 时髦女服商店,时髦女服商/女设计师
coverage ['kʌvəridʒ] *n*. 新闻报导
craft [kræft] *n*. 手艺,技巧,手工;手工艺品行业
creativity [ˌkriːei'tivəti] *n*. 创造力,创造
credibility [ˌkredə'biliti] *n*. 可信程度,确实性
crepe [kreip] *n*. 绉绸;绉布
crisp [krisp] *adj*. 挺括的,挺爽的
criteria [krai'tiəriə] *n*. 标准
criticism ['kritisizəm] *n*. 批评,非难;评论
critique [kri'tiːk] *n*. 评论,评价
crochet hook 钩针
crochet ['krəuʃei] *n*. 钩针,钩针编制品;钩编花边

cropping ['krɔpiŋ] *v*. 剪除,剪去
cross-like 十字形的
cross-section 横截面,剖视图
crossway ['krɔswei] *n*. 斜纹,交叉
crosswise ['krɔsˌwaiz] *adv*. 横向地
crown ease 袖山吃势
crown notch 袖山对位剪口
crown [kraun] *n*. 袖山
crucial ['kruːʃəl] *adj*. 决定性的;关键的,重要的
crucifix ['kruːsəˌfiks] *n*. 耶稣受难像,十字架
crush [krʌʃ] *v*. 压皱,揉皱
cubist shape 立体形象
cuff ['kʌf] *n*. 袖口
cuffed [kʌfd] *adj*. 翻边的,折边的
cultivated silk 家蚕丝
curved [kəːv] *adj*. 曲线的
custom designer 定制服装设计师
customization [kʌstəˌmaɪ'zeʃən] *n*. 用户化,专用化,定制
customize ['kʌstəmaiz] *v*. 定制,按规格改制,定做
cut-and-sew 裁剪成型(相对于全成型针织服装成衣法)
cut-in-one 连裁
cut-path 裁剪轨迹
cutting ['kʌtiŋ] *n*. 裁剪

D

dark color 深色,暗色
dark [daːk] *n*. 暗色,深色
dart [daːt] *n*. 省道,缉省
decentralize [diː'sentrəˌlaiz] *v*. 分散
decision-making 决策,决策的
decorate ['dekəreit] *v*. 装饰
decorative stitch line 装饰缉线
define [di'fain] *v*. 定义,规定
definition [ˌdefi'niʃən] *n*. 定义
definitive [di'finitiv] *adj*. 确定的;明确的
delivery [di'livəri] *n*. 交货,出货
demise [di'maiz] *n*. 结束,完结
democracy [di'mɔkrəsi] *n*. 民主
denim bottoms 粗斜纹劳动布下装,牛仔裤类服装
denim ['denim] *n*. 斜纹粗棉布,牛仔布,劳动布
denote [di'nəut] *v*. 指示,表示
department [di'paːtmənt] *n*. 部门
derive from 由来,起源自……
descending [di'sendiŋ] *adj*. 递降的
description [dis'kripʃən] *n*. 叙述,描写,描绘,说明书
designer wear 设计师(标名)服装
designer [di'zainə] *n*. 设计师
desirable [di'zaiərəbl] *adj*. 理想的,令人满意的,良好的,优良的
detail-oriented 针对细节,细心的
deteriorate [di'tiəriəreit] *v*. 损坏,损耗,变质
deviation [ˌdiːvi'eiʃən] *n*. 偏离,偏向,偏差
diagonal [dai'ægənəl] *adj*. 对角线的,斜的,斜纹的
diagram ['daiəgræm] *n*. 图表
diameter [dai'æmitə] *n*. 直径
dictatorship [dik'teitəʃip] *n*. 主宰,独裁
die [dai] *n*. 刀模
differentiation [ˌdifəˌrenʃi'eiʃən] *n*. 区别,分化,变异
dimensional stability 尺寸稳定性
dinner jacket 小礼服,晚宴茄克礼服,晚礼服
dip into 浸入
discounter ['diskauntə] *n*. 折扣商店
disparate ['dispərit] *adj*. 完全不同的,全异的
display [dis'plei] *n*. 展览,陈列 *v*. 展示,陈列,展出

dissect [di'sekt] *v*. 解剖,切开
disseminate [di'semineit] *v*. 撒播,传播,散布
distinct [dis'tiŋkt] *adj*. 独特的;明显的;不寻常的
distinctive [di'stiŋktiv] *adj*. 明显不同的,特别的,突出的
distinctly [dis'tiŋktli] *adv*. 显然地,明显,清楚地
distribution centre (DC)配送中心
distribution [distri'bjuːʃən] *n*. 分发,分配
diversify [dai'vəːsifai] *adj*. 形形色色的,多种多样的
docket ['dɔkit] *n*. 标签,签条,标志
double-face 双面的
dozens of 许多的
draft [draːft] *n*. 样板,原图,草案
drape [dreip] *v*. 垂坠,悬垂,立体裁剪
dress form 人体模型架,胸架
dress [dres] *v*. 穿着; *n*. 服装,礼服,连衣裙
dressmaker ['dresˌmeikə] *n*. 裁缝工,服装工;女服裁缝师
dry cleaning 干洗
dry-cleaned 干洗
durability [ˌdjurə'biliti] *n*. 耐久性,耐用性,坚固
durable press 耐久压烫
dusty ['dʌstiː] *adj*. 不明朗的,灰暗的
dye [dai] *n*. 颜料,染料; *v*. 染,染色

E

earlobe ['iəˌləʊb] *n*. 耳垂
earth shoe 大地鞋
ease of care 易护理的
ease [iːz] *n*. 松份;放松,放宽;拔开;舒适
easy fitting 穿着舒适的
e-commerce 电子商务
economical [ˌikə'nɔmikəl] *adj*. 节俭的,经济的,合算的
economically [iːkə'nɔmikəli] *adv*. 节约地,不浪费地,节省地
efficiently [i'fiʃəntli] *adj*. 能胜任的,有能力的,效率高的
elaborate [i'læbərət] *adj*. 精致的,精巧的
elastic [i'læstik] *n*. 弹松带,松紧带,橡皮带;弹性织物
elbow line (EL)肘围线
elbow ['elbəu] *n*. 肘部
electronically [ilek'trɔnikli] *adv*. 电子地,电子操纵地
elegant ['eligənt] *adj*. 优雅的,优美的
eliminate [i'limineit] *v*. 除去,排除,删除
elite [i'liːt] *n*. 精英,中坚
embroider [em'brɔidə] *v*. 刺绣,在……上绣花
embroiderer [em'brɔidərə] *n*. 绣花机;绣花工
emerald ['emərəld] *n*. 翡翠,绿宝石
emphasis ['emfəsis] *n*. 强调,重点
emphasize ['emfəsaiz] *v*. 强调
empire waist 帝国式腰线(高腰节)
enclosed seam 封边缝,止口缝
enforce [in'fɔːs] *v*. 加强,强化
engrave [in'greiv] *v*. 雕刻,刻印
enhancement [in'haːnsmənt] *n*. 增强,促进,提高
enlighten [in'laitən] *v*. 启发,开导
enormous amounts of 巨大的,庞大的
entrust [in'trʌst] *v*. 委托,托付
environmental consideration 有关环境方面的考虑
environmental [enˌvaiərən'mentl] *adj*. 周围的,环境的
epitome [i'pitəmi] *n*. 概要,缩影,象征
equivalent [i'kwivələnt] *n*. 同等物,相等物

Erotic Fashion Show　艳情时装表演
establish [is'tæbliʃ] v. 成立,建立,设立;创立;开设,确立
established [is'tæbliʃt] adj. 建立的,固定的,既定的,确定的
ethical ['eθikəl] adj. 伦理的
evaluate [i'væljueit] v. 评估,评价,赋值
evaporate [i'væpəreit] v. 使蒸发;使脱水
even ['iːvən] adj. 平均的,均等的
evening-wear　夜礼服
evolve [i'vɔlv] v. 展开;使发展
exaggerate [ig'zædʒəreit] v. 夸大,夸张,使过大
exclusive [iks'kluːsiv] adj. 唯一的,排他的
execute ['eksikjuːt] v. 执行,实行,完成
execution [ˌeksi'kjuːʃən] n. 执行,实行
executive [ig'zekjutiv] n. 执行部门,执行者,经理主管人员　adj. 行政的
expand [iks'pænd] v. 扩大,扩展,扩张
expectation [ˌekspek'teiʃən] n. 期待,指望,展望
experimental [iksˌperi'mentəl] adj. 实验的;经验的
expert ['ekspəːt] n. 专家,内行,能手
exploitation [ˌeksplɔi'teiʃən] n. 开发,开采
exposure [iks'pəuʒə] v. 暴露,揭发,揭露
extension [iks'tenʃən] v. 延长,扩充,范围,扩展名
extent [iks'tent] n. 范围,程度
external [eks'təːnl] adj. 外部的,外在的
extraordinarily [iks'trɔːdinərili] adj. 非凡的,特别的,非常的,使人惊奇的
extremely [iks'triːmli] adv. 极端地,非常地

F

fabric ['fæbrik] n. 织物,织品,面料,布
fabrication [ˌfæbri'keiʃən] n. 制造,生产,制作,加工;成品
facilitate [fə'siliteit] v. 使……容易,使不费力,促进,帮助
facing ['feisiŋ] n. 挂面,贴条,贴边
fad [fæd] n. 流行一时的服装;流行快潮;一时的风尚
fade [feid] v. 褪色,枯萎;凋谢
fair trade　公平贸易,互惠贸易
fair [fεə] n. 博览会,商品展览会,商品交易会
fashion consultant　时装顾问,服饰顾问
fashion curve　流行的周期曲线
fashion cycle　时装界
fashion illustrator　时装画
fashionability ['fæʃənəbliti] n. 时尚,流行
fashionable ['fæʃənəbl] n. 时髦人物,流行的 adj. 时尚的
fashion-conscious　有时尚意识的
fashioned knitting machine　成型的针织机
fashionist ['fæʃənist] n. 时尚追随者;时装店主
fashion-show　时装展览,时装表演
fastener ['faːsənə] n. 紧固件
fastening ['faːsniŋ] n. 扣合件
feature in　占重要位置
fee-based　以收费为基础
feeding ['fiːdiŋ] n. 推布,送料
felt [felt] n. 毛毯,毡
femininity [femi'niniti] n. 女性气质,女人味
fetish ['fetiʃ] n. 偶像
fiber ['faibə] n. 纤维
fifth-scale block　1∶5 纸样
figure ['figə] n. 外形,形状,体形
filament ['filəmənt] n. 长丝
filling ['filiŋ] n. 纬纱
financial [fai'nænʃəl] adj. 财政的,财务的,金融的

finding [ˈfaindiŋ] *n*. 服装附件
fine [fain] *adj*. 细的
fine-tune　微调
finish [ˈfiniʃ] *n*. 织物整理,后整理
firmly [ˈfəːmli] *adv*. 厚实地,坚实地,有身骨
fish-eye dart　鱼眼形省道
fit [fit] *v*. 合身,合体,使合身
flame proofing　抗燃,阻燃,防火
flap [flæp] *n*. 前门襟;袋盖;帽边
flat pattern　平面纸样
flat [flæt] *n*. 平面图,效果图
flatten [ˈflætn] *v*. 拉平
flatter [ˈflætə] *v*. (画像等的形象)美于(真人[实物])
flax [flæks] *n*. 亚麻,麻布
fleece [fliːs] *n*. 羊毛,似羊毛物,羊毛标签
flexible [ˈfleksəbl] *adj*. 弯曲的,灵活的,弹力的
flocked fabric　植绒织物
flora [ˈflɔːrə] *n*. 花卉,植物
focal point　焦点
folding [ˈfəʊldiŋ] *n*. 折叠
foldline　折边线,对折线,折叠线
footwear [ˈfʊtˌweə] *n*. 鞋袜(统称),鞋类(总称)
forecaster [ˈfɔːkaːstə] *n*. 预测者
forefront [ˈfɔːˌfrʌnt] *n*. 最前部, 最前沿
foremost [ˈfɔːməust] *adj*. 最初的,最先的,第一流的
form [fɔːm] *n*. 造型,外形,体型;人体模型
formal [ˈfɔːməl] *adj*. 正式的
format [ˈfɔːmæt] *n*. 开本,版式,形式,格式
Fr.　from 的缩写
fragment [ˈfrægmənt] *n*. 碎片,片段
frame [freim] *n*. 结构,框架
fray [frei] *v*. 磨损
frequently [ˈfriːkwəntli] *adj*. 频繁的,常见的
fringe [frindʒ] *n*. 毛边,蓬边;流苏
front bodice　前衣身
front central line　(FCL)前中心线
front neck point　(FNP)前颈窝点
front pitch　前袖窿吻合点
Frt.　front 的缩写
fruitful sales　富有成果的销售
full skirt　宽裙,宽下摆群,喇叭长裙,整圆裙
fullness [ˈfulnis] *n*. 丰满度
full-scale blocks　原尺寸,实际尺寸
full-size　原尺寸的,实际尺寸的
fully-fashioning　全成形的
functional [ˈfʌŋkʃənl] *adj*. 机能的,功能的
funeral [ˈfjuːnərəl] *n*. 葬礼
fur [fəː] *n*. 毛皮
fuse [fjuːz] *v*. 熔化,使融合,合并

G

garment [ˈgaːmənt] *n*. 衣服,服装
gather [ˈgæðə] *n*. 褶裥,碎褶　*v*. 打褶裥
gathering foot　碎褶压脚
gauge [geidʒ] *n*. 标准规格,标准尺寸
gauntlet [ˈgɔːntlit] *n*. 长手套,防护手套
gear toward　专门从事某事
gear to　适合于
gender [ˈdʒendə] *n*. 性别
geometrics [ˌdʒiːəˈmetriks] *n*. 几何图形
giant [ˈdʒaiənt] *adj*. 庞大的,巨大的
glamorous [ˈglæmərəs] *adj*. 迷人的
glaze [gleiz] *v*. 上光,极光;光泽,色泽
Glissade [gliˈsaːd] *n*. 格利萨特里子布
glitz [glits] *adj*. 闪光的,炫目
globally [ˈgləubəl] *adj*. 全球的
glossy [ˈglɔːsiː, ˈglɔsi] *adj*. 有光泽的,光鲜的
glue [gluː] *v*. 胶合
godet [gəuˈdet] *n*. 三角布(用以放宽衣裙

下摆),裆布
gore [gɔː] *n*. 衣片,拼块,三角布
gothic clothing 哥特式服装
grade rule 放码规则,推档规则
graduation [grædju'eiʃən] *n*. 毕业典礼
grain line 经向线,经向标志
graphite paper 石墨纸
gray-blue 灰蓝色
greed-upon 互相认可的
grip [grip] *n*. 夹具,夹子 *v*. 夹紧,抓住,握住
grommet holes 金属孔眼,索环
groom [grum] *n*. 新郎
groundwork ['graundwəːk] *n*. 基础
guarantee [ˌgærən'tiː] *v*. 保证,担保; *n*. 保证书,保证人
guideline ['gaidlain] *n*. 指导方针
gusset ['gʌsit] *n*. 三角形衬料

H

hair style 发式,发型
hairbrush ['heəˌbrʌʃ] *n*. 发刷
hand wash 手洗
hand [hænd] *n*. 手感
handbag ['hændbæg] *n*. 手袋
hand-crafted 手工制作的
hand-cut 手工裁剪
hand-in-hand 手牵手的, 亲密的;并进的
handle ['hændl] *n*. 手感
hangtag ['hæŋtæg] *n*. 使用说明,飘带式商品标签,吊牌
hank silk 丝线团
hard wear 经穿,耐磨
hardened ['haːdənd] *adj*. 坚固的,坚硬的
hard-to-reach 难以达到的,难以触及的
hat [hæt] *n*. 帽子
haute couture 高级女式时装,高级定制
head size (HS)头围
heat-sealed 热封的
heat-setting 热定型
heavy-duty 耐用的,重型的
heavyset ['hevi'set] *adj*. 体格魁伟的
heavy-weight 厚重
height [hait] *n*. 身高
hem [hem] *n*. 贴边,卷边,下摆,脚口折边,缝边下缘
herringbone structure 人字形结构
High TOL 上误差
high-priced 高档价格
high-quality 高级的
high-resolution 高分辨率,高清晰度
hip line (HL)臀围线
hipline ['hiplain] *n*. 臀围,(女裙的)臀围部分
historian [his'tɔːriən] *n*. 历史学家
hockey ['hɔki] *n*. 曲棍球,冰球
hooks and eyes 钩眼扣子;风钩
horizontal [ˌhɔri'zɔntəl] *adj*. 地平的,地平线的,水平的
hosiery ['həuziəri] *n*. 男袜,男用针织品
hot knife 热封刀,热封钳
hue [hjuː] *n*. 颜色,色彩,色相
hygiene product 卫生用品

I

identify [ai'dentifai] *v*. 认出;识别;鉴别;验明
illuminate [i'ljuːmineit] *v*. 照亮
image ['imidʒ] *n*. 影像,肖像,图像,形象,反映
imitate ['imiteit] *v*. 模仿;模拟;仿效;效法,仿造
imitative ['imitətiv] *adj*. 模仿的,仿效的
impart [im'paːt] *v*. 给予,传授,告知
imperative [im'perətiv] *adj*. 绝对必要的,不可避免的

imperfection [ˌimpəˈfekʃən] *n*. 不完整性，不足；缺点

in conjunction with　连合，结合，联系

in earnest　认真地，诚挚地

in harmony with　与……协调，与……一致

in terms of　用……的话，根据，按照

inconsistent [ˌinkənˈsistənt] *adj*. 断断续续的，不连贯的

increasingly [inˈkriːsiŋli] *adv*. 逐渐地，渐增地

incredible [inˈkredəbl] *adj*. 难以置信的，不可思议的，惊人的

increment [ˈinkrimənt] *n*. 档差

independent testing labs　独立的测试实验室

indicate [ˈindikeit] *v*. 指示；指出，表明；显示；象征

indispensable [ˌindisˈpensəbl] *adj*. 不可缺少，绝对必要的

indisputably [ˌindiˈspjuːtəbli] *adv*. 无可争辩，无可置疑

individual [ˌindiˈvidjuəl] *adj*. 个人的，个别的，单独的，个性的

individualist [ˌindiˈvidjuəlist] *n*. 个人

individuality [indiˌvidjuˈæliti] *n*. 个体，个性

indulgence [inˈdʌldʒəns] *v*. 沉迷，沉溺

industrial pattern　工业用纸样

industrialize [inˈdʌstriəlaiz] *v*. 使工业化，实现工业化

industry standard　行业标准

inexpensive [ˌiniksˈpensiv] *adj*. 廉价的，便宜的

infant [ˈinfənt] *n*. 婴儿，幼童

inferential [ˌinfəˈrenʃəl] *adj*. 推理的，可以推论的

influx [ˈinflʌks] *v*. 汇集；涌进，涌入

inform [inˈfɔːm] *v*. 通知，告诉

in-house　公司内部的，内部的，室内的

initial phases　初始阶段

initial [iˈniʃəl] *adj*. 最初的，初期的；*n*. 姓名的开头字母

injunction [inˈdʒʌŋkʃən] *n*. 命令，禁令

inlay [ˈinlei] *v*. 镶嵌，嵌 [插] 入

innovation [ˌinəuˈveiʃən] *n*. 创新，革新，改革，

innovative [ˈinəuveitiv] *adj*. 革新的，创新的，富有革新精神的

innovator [ˈinəveitə] *n*. 改革者，革新者

inspect [inˈspekt] *v*. 检阅，检查，审查，视察

inspiration [ˌinspəˈreiʃən] *n*. 灵感

instill [inˈstil] *v*. 滴注，慢慢地灌输

instruction book　指导书，参考书

instruction [inˈstrʌkʃən] *n*. 说明书，细则

insulate [ˈinsjuleit] *v*. 隔离，使孤立

intarsia [inˈtaːsiə] *n*. 嵌花，镶嵌装饰

integral [ˈintigrəl] *adj*. 组成的，必备的

integrity [inˈtegriti] *n*. 诚实，正直，廉正

intended [inˈtendid] *adj*. 有意的，故意的

intensely [inˈtensli] *adv*. 强烈地

intensity [inˈtensiti] *n*. 强烈

interfacing [ˈintəˌfeisiŋ] *n*. 粘合衬

interlace [ˌintəˈleis] *v*. 交织

interlining [ˈintəˈlainiŋ] *n*. 衣服衬里，衣服衬里的布料

interlock [ˌintəˈlɔk] *v*. 联锁，双罗纹，使连接

internal [inˈtəːnəl] *adj*. 内部的，内在的

internationally [ˌintəˈnæʃənəli] *adv*. 国际性地，在国际间

internet-based　基于网络的

interrelationship [ˈintə(ː)riˈleiʃənʃip] *n*. 相互关系

interview [ˈintəvjuː] *n*. 面谈，访问，接见，面试

interweave [ˌintəˈwiːv] *v*. 交织，混合

intricate ['intrikit] *adj*. 复杂的,缠结的
intrinsic characteristic　内在特性
inventory level　存货水准,库存水平
inventory software　资产管理软件
inventory ['invəntəri] *n*. 财产等的清单;商品的目录;任何详细记载
investigatory [in'vestigeitəri] *adj*. 研究的,好研究的
invisible zipper foot　暗拉链压脚
involvement [in'vɔlvmənt] *n*. 卷入,介入
iris ['aiəris]　鸢尾花
iron ['aiən] *n*. 烙铁,熨斗
ironing board　熨衣板
ironing ['aiəniŋ] *n*. 熨烫
irreplaceable [ˌiri'pleisəbl] *adj*. 不能替代的
Islamic headscarf　伊斯兰头巾
issue ['isjuː] *v*. 颁布,发布,发行
item ['aitem] *n*. 项;条款;项目;产品;展品
ivory ['aivəri] *n*. 象牙;(海象等的)长牙

J

jacket blazer　运动夹克
jacquard [dʒə'kaːd] *n*. 提花机,提花织物
Javanese sarong　爪哇莎笼围裙
jaw [dʒɔː] *n*. 夹片,虎钳牙
jeans [dʒiːnz] *n*. 牛仔裤,紧身裤,粗斜纹棉布裤
jersey suit　针织套装
jersey ['dʒəːzi] *n*. 紧身运动套衫;平针织物
jersey-wear　针织物,针织坯布;弹力针织物
jewelry ['dʒuːəlriː] *n*. 首饰,珠宝饰物
jigsaw puzzle　拼图玩具
jobber ['dʒɔbə] *n*. 批发,批发商
junior tops　少年上装
junior ['dʒuːnjə] *n*. 少年,瘦小的女服尺寸,少女型
justify ['dʒʌstifai] *v*. 证明……有道理,为……辩护

K

keep track of　与……保持联系
kimono [kə'məʊnə] *n*. 连袖;连袖服装;和服;和服式女晨衣
knee line　(KL)膝围线
knit top　针织上衣
knit [nit] *n*. 针织,针织品,针织服装　*v*. 针织
knitted rib　针织罗纹
knitwear ['nitˌweə] *n*. 针织品,针织衣物
knock off　翻制设计,翻印本
knotting ['nɔtiŋ] *n*. 编结

L

label ['leibəl] *n*. 标号,标签,标记,商标;服装商店(或时装设计师)的标记
labor-intensive　劳动密集型的
lace [leis] *n*. 网眼织物,网眼花边,滚带,饰带
lace-making　花边编织
lacis ['leisis] *n*. 方网眼花边
laminates ['læməˌneitz] *n*. 粘合布,多层(粘合)布
lamination [ˌlæmi'neiʃən] *n*. 粘合衬
laser ['leizə] *n*. 激光
latch [lætʃ] *v*. 获得,缠住,占有
latter ['lætə] *adj*. 后者的
launder ['lɔːndə] *v*. 洗 [烫]衣
lavishly ['læviʃli] *adv*. 丰富地
lay plan　铺料
lay [lei] *v*. 铺放,铺料
leather ['leðə] *n*. 皮革,皮革制品
left and right-handed　左右手的
leg warmer　暖腿套,护腿
leggings ['legiŋz] *n*. 儿童护腿套裤,开裆

裤,裤绑腿,袜统
legislation [ˌledʒisˈleiʃən] *n*. 法规
legs [legs] *n*. 裤脚
lever [ˈliːvə] *n*. 杠杆
liaise [liːˈeiz] *v*. 与建立……联系
license [ˈlaisəns] *n*. 许可证
lifestyle [ˈlaifˌstail] *n*. 生活方式
light [lait] *n*. 淡色,浅色;光,光线;光泽
light-weight 轻量
ligne 莱尼(钮扣规格:1莱尼=0.633毫米)
lime-colored 酸橙绿色的
limited-edition 限量版,限量发行的
line [lain] *n*. 型,款式,纹路;线条,轮廓;系列;衬里
linear trim 线形饰边
linen thread 麻线
linen [ˈlinin] *n*. 亚麻,亚麻制品
lingerie [ˌlaːnʒəˈrei] *n*. 女式内衣
lining [ˈlainiŋ] *n*. 衬里,里子,衬料
link to 把……和……联系起来
literally [ˈlitərəli] *adv*. 按照字面上地
locket [ˈlɔkit] (装有照片或贵重金属纪念品的)项链
lockstitch [ˈlɔkstitʃ] *n*. 锁式线迹;锁缝
lofty [ˈlɔfti] *adj*. 膨松的,弹性
logo [ˈlɔgəu] *n*. 图形,商标,图样
look [luk] *n*. 风貌,风格、型;款式,外表,姿态
loop fabric 毛圈织物,毛巾织物
loop turner 翻带器
loop [luːp] *n*. 线环,布环,毛圈
Low TOL 下误差
low-luster 光泽暗淡的,无光泽
loyalty [ˈlɔiəlti] *n*. 忠诚
luster [ˈlʌstə] *n*. 光泽,光彩
lustrous [ˈlʌstrəs] *adj*. 光亮的,鲜艳的
luxury [ˈlʌkʃəri] *n*. 奢侈,豪华,奢侈品
lyocell [ˈliəˌsel] *n*. 天丝,莱塞尔纤维

M

machine cotton and silk 缝纫机用棉线,丝线
machine-stitched 机缝的
machinist [məˈʃiːnist] *n*. 机械师,机械修理工
macrame [məˈkraːmi] *n*. 流苏,花边
madras [məˈdrɑːs] *n*. 马德拉斯狭条衬衫布
mainstay [ˈmeinˌstei] *n*. 支柱
mainstream [ˈmeinˌstriːm] *n*. 主流
maintenance [ˈmeintinəns] *v*. 维持,保持
makeup [ˈmeikˌʌp] *n*. 化妆;化妆品
mandatory standard 法定标准
mandatory [ˈmændətəri] *adj*. 受委托的,强迫的,强制的
manipulate [məˈnipjuleit] *v*. 操纵,利用,操作,巧妙地处理
man-made 人造的,合成的
mannequin [ˈmænikin] *n*. 服装模特儿;(橱窗里的)服装模型;人体模型
manner [ˈmænə] *n*. 方式,方法;态度
manual [ˈmænjuəl] *adj*. 手动的,手工的
marital status 婚姻状态
mark [maːk] *n*. 标记,标志,记号,符号,型号 *v*. 标记,标示
markdown [ˈmaːkdaun] *n*. 减价商品,削价商品
marker-making system (服装)排料系统
mass market 大众市场;大规模市场
mass merchandiser 超型市场
master pattern 基本纸样,母板
mastery [ˈmaːstəri] *n*. 精通,掌握
matching [ˈmætʃiŋ] *adj*. 相配的
matellasse [matˈlɑː sa] *n*. 马特拉塞凸纹布
matte [mæt] *n*. 暗淡色调
mauve [məʊv] *adj*. 紫红色(的);淡紫色(的)
maximum [ˈmæksiməm] *n*. 极点,最大量,

极大

meaningful [ˈmiːniŋful] *adj*. 意味深长的，很有意义的

measurable [ˈmeʒərəbl] *adj*. 可量的，可测量

measure base 测量单位

measurement [ˈmeʒəmənt] *n*. 围度，测量，尺寸

mechanize [ˈmekənaiz] *v*. 机械化，机械化（生产）

medium-weight 中厚

membrane [ˈmemˌbrein] *n*. 膜，薄膜

mending [ˈmendiŋ] *n*. 织补，缝补；修理

menswear [ˈmenzˌweə] *n*. 男服

microprocessor [ˈmaikrəuprəusesə] *n*. 微处理机

middle hip line (MHL)中臀围线

military suit 军装

mill [mil] *v*. 缩绒，缩呢

millimeter [ˈmilimiːtə] *n*. 毫米

millinery [ˈmiləˌneriː] *n*. 女帽，妇女头饰，女帽制造商，女帽商

mimic [ˈmimik] *n*. 模仿，仿制品

mineral [ˈminərəl] *n*. 矿物，无机物

minimize [ˈminimaiz] *v*. 将……减到最少，最小化

minimum [ˈminiməm] *n*. 最小(量)，最低额，最低点

miniskirt [ˈminiˌskəːt] *n*. 超短裙，迷你裙

mink coat 貂皮大衣

minority [maiˈnɔriti] *adj*. 少数，少数的

miscellaneous [misiˈleinjəs] *adj*. 各种的，多方面的

mitre [ˈmaitə] *n*. 45°折角拼缝，斜拼接，斜接缝

mitten [ˈmitn] *n*. 连指手套

mixture [ˈmikstʃə] *v*. 混合，混合状态；*n*. 混合物

mode agency 模特介绍所，模特经纪所

mode [məud] *n*. (服式的)式样，风尚，风气，流行

model [ˈmɔdəl] *n*. 型，型号；时装模特，商品模特；款式，模型

modernity [mɔˈdəːniti] *n*. 现代性，现代状态，现代东西

modification [ˌmɔdəfiˈkeiʃən] *n*. 修改，修正

modify [ˈmɔdifai] *v*. 修正，修改

moisture absorbency 吸湿性

moisture [ˈmɔistʃə] *n*. 湿气，水分，潮湿；水蒸气

molding [ˈməuldiŋ] *n*. 模制，浇铸

monogram [ˈmɔnəugræm] *n*. 字母组合，交织字母

mosaic [mɔˈzeiik] *n*. 马赛克，马赛克(图案)

moth [mɔθ] *n*. 蛾，蛀虫

motif [məʊˈtiːf] *n*. 基调，基本图案，基本色调

multi-layer 多层

mummy [ˈmʌmiː] *n*. 木乃伊

mundane [mʌnˈdein] *adj*. 现世的，世俗的

muslin [ˈmʌzlin] *n*. 平纹细布，麦斯林纱

muted [ˈmjuːtid] *adj*. 晕的；模糊的；暗淡的；柔和的

N

nap [næp] *n*. 衣料起毛 *v*. 使衣料起毛

nape [neip] *n*. 颈背，后颈

napped fabric 起绒面料

narrow hemmer foot 狭卷边压脚

neat [niːt] *adj*. 整洁的，匀称的

neck drop 领口深

neck [nek] *n*. 颈部，领部

necklace [ˈneklis] *n*. 项链

necklace line (NL)颈围线

neckline [ˈneklain] *n*. 领围线

necktie [ˈnekˌtai] *n*. 领带,领结

needle or velvet board (供绒毛织物整烫的)针毯烫垫

needlepoint [ˈniːdlpɔint] *n*. 刺绣品;针绣

negotiate [niˈgəuʃieit] *v*. 议定,商定,谈判

negotiation [niˌgəuʃiːˈeiʃən] *n*. 协定,交易,协商,谈判

nested pieces 嵌套式样片

net [net] *n*. 网眼织物

net pattern 净样

neutral [ˈnjuːtrəl] *n*. 中和色,不鲜明的颜色,与灰色相协调的颜色

nevertheless [ˌnevəðəˈles] *adv*. 然而,虽然如此

niche [nitʃ] *n*. 适当的位置,恰当的处所

nightwear [ˈnaitweə] *n*. 晚服,夜间家常服

nongovernmental [ˈnɔnˈgʌvənməntəl] *adj*. 非政府的,非政治(上)的

nonverbal [ˈnɔnˈvəːbəl] *adj*. 非语言的

non-woven 非织物,无纺的

notable [ˈnəutəbl] *n*. 著名人士

notch [nɔtʃ] *n*. 刀口,刀眼 *v*. 打刀眼

novel [ˈnɔvəl] *adj*. 新奇的,新颖的

numerous [ˈnjuːmərəs] *adj*. 许多的,无数的

nylon [ˈnailən] *n*. 尼龙,酰胺纤维;尼龙制品

O

observation [ˌɔbzəːˈveiʃən] *n*. 注意,观察

occasion-wear 应时服装;特定场合服装

odd [ɔd] *adj*. 奇特的,古怪的

odd-shaped 不规则形状的

off-center 偏离中心的(地)

old-fashioned 过时的

olefin [ˈəuləfin] *n*. 烯烃

on-center 居中

one-of-a-kind 独一无二的

one-way 一顺的,单向的

ongoing [ˈɔnˌgəuiŋ] *adj*. 进行的

on-site 现场的

opening [ˈəupəniŋ] *n*. 开襟,开门

open-to-buy (OTB)进货限额,许购定额

open-weave 稀薄组织

openwork [ˈəupənwəːk] *n*. 网状织物,网眼式

opportunity [ˌɔpəˈtjuːniti] *n*. 机会,时机

optimal [ˈɔptiməl] *adj*. 最佳的,最理想的

ordinary [ˈɔːdinəri] *adj*. 原始的,普通的,平凡的

organic fiber 有机纤维

orientation [ˌɔːrienˈteiʃən] *n*. 方向,目标

originate [əˈridʒineit] *v*. 产生,引起

outer [ˈautə] *adj*. 外在的,外部的

outerwear [ˈautəweə] *n*. 外衣,外套,户外穿服装

outfit [ˈautfit] *n*. 服装,全套服装

outward [ˈautwəd] *adj*. 外面的,明显的,公开的

overall [ˈəuvərɔːl] *n*. 宽大罩衫,工作罩衣

over-buy 溢价购买

overcast [ˈəuvəkaːst] *adj*. 覆盖的,遮盖的,包边缝纫

over-casting 包边,锁边

over-garment 大衣,罩袍

overlapping [ˈəuvəˈlæpiŋ] *n*. 重叠,交叠

over-locker 拷边机,包缝机

over-purchase 溢价购置

oversize [ˈəuvəˈsaiz] *n*. 特大型,特大尺寸,尺寸过大

P

pad [pæd] *n*. 衬垫,肩垫

padded shoulders 垫肩

painting [ˈpeintiŋ] *n*. 手绘

pair with 与……成对

panel [ˈpænəl] *n*. 嵌条,饰条,布块
pants [pænts] *n*. 裤子,长裤,便裤
parameter [pəˈræmitə] *n*. 参数,参量
paramount [ˈpærəˌmaʊnt] *adj*. 最重要的,最高的
paratrooper pant 伞兵裤
Parisian [pəˈrizjən] *adj*. 巴黎的,巴黎人的
participant [pa:ˈtisipənt] *n*. 参加者
participation [pa:ˌtisiˈpeiʃən] *v*. 参加,参与
passementerie [pasˈmətri:] *n*. 边饰,金银线镶边,珠饰
patchwork [ˈpætʃwə:k] *n*. 拼缝;拼缝品
pattern generation software 纸样生成(绘制)软件
pattern maker 样板师,纸板师,打样师
pattern [ˈpætən] *n*. 纸样,裁剪样板;花纹组织;图案;花样,式样
payment term 支付条款
peak [pi:k] *n*. 巅,顶点;到达最高点
peel off 去皮,剥离
pendant [ˈpendənt]垂饰,挂件;有垂饰的项链
pendulum [ˈpendjuləm] *n*. 钟摆
penetrate [ˈpenitreit] *v*. 穿透,穿过,透过
peplum [ˈpepləm] *n*. 腰褶
percentage [pəˈsentidʒ] *n*. 百分比,比率,部分,可能性
performance [pəˈfɔ:məns] *n*. 履行,实行,执行,完成
perfume [ˈpə:fju:m] *n*. 香料,香水
permanent [ˈpə:mənənt] *adj*. 永久的
perseverance [ˌpə:siˈviərəns] *n*. 坚持
personalities [ˌpə:səˈnælitiz] *n*. 知名人士,名流
personality [pə:səˈnæliti] *n*. 人的存在;个性,人格
personalization [ˌpə:sənəlaiˈzeʃən] *n*. 个性化
personnel [ˌpə:səˈnel] *n*. 人员,职员
person-to-person 个人之间的,面对面的
perspiration [ˌpə:spəˈreiʃən] *n*. 汗(水);出汗
pertinent [ˈpə:tinənt] *adj*. 有关的,相关的
petroleum [piˈtrəuliəm] *n*. 石油
philosophy [fiˈlɔsəfi] *n*. 基本原理;见解
photogenic [ˌfəutəuˈdʒenik] *adj*. 适宜于拍照的,拍照效果好的,特别上镜的
piece goods 匹头,布匹
piece [pi:s] *n*. 衣片,裁片;部分;匹;件
pierce [piəs] *v*. 刺入;刺穿;穿透
pile fabric 绒毛织物,起绒织物,割绒织物
pile [pail] *n*. 绒毛,绒头,毛茸
pill [pil] *v*. 起球
pillage [ˈpilidʒ] *v*. 起球 *n*. 起球
pin [pin] *n*. 大头针,别针
pinking shears 锯齿剪,花边剪
pipeline [ˈpaipˌlain] *n*. 流水线,商品供应线
piping foot 滚边压脚
pivotal [ˈpivətl] *adj*. 关键性的
placket [ˈplækit] *n*. 开口,衣袋
plaid [plæd] *n*. 方格布,方格呢;格子花纹
play out 放出,用完,结束
pleat line 裥位线
pleat [pli:t] *n*. 打褶,褶裥
pleated trouser 有褶裤
plot [plɔt] *v*. 绘制平面图
plotter [ˈplɔtə] *n*. 描绘器,图形显示器,绘图器;标图员
pluck [ˈplʌk] *v*. 拔毛,拉毛
ply [plai] *n*. 织物层数
pocket mouth 袋口
pocket placing 袋位
pocket [ˈpɔkit] *n*. 口袋
point presser 马凳,烫凳,小烫台,小烫板
point-of-purchase 卖点
polyester [ˌpɔliˈestə(r)] *n*. 聚酯

polyurethane [ˌpɔliˈjuəriθein] *n*. 聚氨基甲酸脂;聚氨酯(类)
Poor Look　破旧型款式,贫穷装
pop art　通俗艺术,波普艺术
popular [ˈpɔpjulə] *adj*. 流行的,普及的
popularity [ˌpɔpjuˈlæritiː] *n*. 流行,普及
popularize [ˈpɔpjuləraiz] *v*. 推广
portable [ˈpɔːtəbl] *adj*. 便于携带的;手提式的,轻便的,可移动的
pose [pəuz] *n*. 造型,姿态
potential [pəˈtenʃəl] *adj*. 潜在的
potter [ˈpɔtə] *n*. 陶工,制陶工人
power loom　动力织布机
practical [ˈpræktikəl] *adj*. 实用的
precise [priˈsais] *adj*. 精确的,明确的
precision [priˈsiʒən] *n*. 精密,精确
preference [ˈprefərəns] *n*. 偏爱,喜爱
prehistoric [ˈpriːhisˈtɔrik] *adj*. 史前的
preliminary stage　初级阶段
preliminary [priˈliminəri] *adj*. 预备的,开端的,初步的,开端
preproduction [priːprəˈdʌkʃən] *adj*. 试生产的
prespecified [priˈspesifaid] *adj*. 预定的,预先设计好的
press cloth　熨烫衬布,熨烫垫布
press mitt　手套式烫垫
press [pres] *n*. 出版社,出版机构;记者;新闻舆论
pressing [ˈpresiŋ] *n*. 熨烫,压呢,烫衣
prestigious [presˈtiːdʒəs] *adj*. 有威望,有声誉的
Prêt-a-Porter　现成的服装,高级女装成衣
pretty [ˈpriti] *adj*. 优雅的,潇洒的
princess line　公主线
princess seams　公主线
principal [ˈprinsəpəl] *adj*. 主要的,最重要的,首要的
prints [print] *n*. 印花,印花布
prior to　在前,居先,在……之前
prism shape　棱镜形状
private label　商店标签;独立设计师标签
procedure [prəˈsiːdʒə] *n*. 过程,步骤程序
prodigally [ˈprɔdigli] *adv*. 浪费地
production-line　生产线,流水线
professional buyer　专业采购员,专业买手
professional [prəˈfeʃənəl] *n*. 自由职业者,专业人士内行,专家
proliferation [prəuˌlifəˈreiʃən] *n*. 增生,扩散
promotion [prəˈməuʃən] *n*. 推广,推销,促销
promotional [prəˈməuʃənl] *adj*. 促销的,推销的
property [ˈprɔpəti] *n*. 财产,所有权,性质,属性
proportion [prəˈpɔːʃən] *n*. 比,比率均衡,相称
prototype [ˈprəutətaip] *n*. 原型,模型,样板;标样,样品
proximity [prɔkˈsimiti] *n*. 接近,邻近
Pt. point 的缩写
publication [ˌpʌbliˈkeiʃən] *n*. 出版物;出版
pucker [ˈpʌkə] *v*. 折叠,起皱
puff [pʌf] *n*. 尖角布;整片;皱褶;胖褶
punch hole　孔位记号;穿孔
punk-rock　朋克～摇滚
purchaser [ˈpəːtʃəsə] *n*. 买方,购买者
purse [pəːs] *n*. 小钱袋,女用小包
PVC(polyvinyl chloride)　聚氯乙烯

Q

qualitative [ˈkwɔlitətiv] *adj*. 性质的,质的,定性的
quality manual　质量手册
quantitative [ˈkwɔntitətiv] *adj*. 数量的,定量的

question-and-answer　问答式
quilt [kwilt] *n*. 被子　*v*. 绗缝
quilted effect　绗缝效果
quilting [ˈkwiltiŋ] *n*. 绗缝

R

raglan [ˈræglən] *n*. 插肩，插肩袖；套袖大衣，插肩袖大衣，连肩，包肩
raincoat [ˈreinkəut] *n*. 雨衣，风雨衣
ratio [ˈreiʃiəu] *n*. 比；比例
raw edge　毛边
raw material　原材料
rayon viscose　粘胶人造丝
rayon [ˈreiɔn] *n*. 人造丝织物，人造丝
ready-to-wear　现成的服装，高级女装成衣
real-time　实时的，快速的
recognition [ˌrekəgˈniʃən] *n*. 认识，识别
reconstituted [ˌriːˈkɔnstitjuːtid] *adj*. 再造的，再生的
reconstruct [ˈriːkənˈstrʌkt] *v*. 重现，再现
recreation [rekriˈeiʃ(ə)n] *n*. 娱乐，消遣
rectangle [ˈrektæŋgl] *n*. 矩形，长方形
reel [riːl] *n*. 线轴
reenactment [riːiˈnæktmənt] *n*. 重演
regenerated [riˈdʒenəˌreitid] *adj*. 再生的
registration [ˌredʒisˈtreiʃən] *n*. 登记
regulate [ˈregjuleit] *v*. 管理，控制
rejection [riˈdʒekʃən] *n*. 抛弃
reliability [riˌlaiəˈbiliti] *n*. 可靠性，安全性；可信赖性
remainder [riˈmeində] *n*. 剩余物，其余的人，余数
remnant [ˈremnənt] *n*. 剩余，零料
render [ˈrendə] *v*. 表现，反映
reorder [ˈriːˈɔːdə] *v*. 翻单，追加订货
replica [ˈreplikə] *n*. 副本
represent [riːpriˈzent] *v*. 表现，表示，代表
representative [ˌrepriˈzentətiv] *n*. 代理
resilience [riˈziljəns] *n*. 回弹，回弹性能
resist bending　抗弯曲
responsibility [riˌspɔnsəˈbiliti] *n*. 任务；职责
restrict [risˈtrikt] *v*. 限制，限定，约束
retailer [riˈteilə] *n*. 零售商
rever [ˈrɪvə] *n*. 翻边，翻领，翻袖
reversible [riˈvəːsəbl] *n*. 双面织物
rhinestone [ˈrainˌstəʊn] *n*. 人造钻石
rib [rib] *n*. 罗纹
ribbon [ˈribən] *n*. 带，丝带，缎带，饰带
rickrack [rikˈræk] *n*. 波曲形花边，荷叶边
rigidly [ˈridʒidli] *adj*. 严格的
rim [rim] *n*. 边，缘；轮缘
rip out　拆除，拆开
rise [raiz] *n*. 立裆，直裆，裆
ritual [ˈritjuəl] *n*. 仪式，典礼
roll [rəul] *v*. 卷成，卷拢，卷起
roller foot　双边固定轮压脚
rotary [ˈrəutəri] *adj*. 旋转的，轮转的
rotate [rəuˈteit] *v*. 旋转；循环，自转轮换
rough [rʌf] *adj*. 布面毛糙的
royalty [ˈrɔiəlti] *n*. 皇室，王族成员，特权阶层
rub [rʌb] *v*. （织物的）擦伤痕
rubber boots　橡胶靴，橡胶长筒靴
ruby [ˈruːbi] *n*. 红宝石
ruler [ˈruːlə] *n*. 直尺
runway [ˈrʌnwei] *n*. 时装天桥

S

safari [səˈfaːriː] *n*. 瑟法里式，猎装，淡土黄色
sake [seik] *n*. 目的，缘故，理由
sale rack　展销，促销
sales sheets　销售（报）表
salespeople [ˈseilzˌpiːpl] *n*. 售货员，店员
sample dimensions　样衣尺寸

sample maker　样衣制作师
sample-making　样品制作
sapphire [ˈsæfˌaiə] *n*. 蓝宝石,人造白宝石
satin [ˈsætn] *n*. 缎子,绸缎
saturate [ˈsætʃəreit] *v*. 渗透,浸透,充满,使饱和
saturation [ˌsætʃəˈreiʃən] *n*. 色品度;纯度,纯色性;饱和度;浓度
saw-tooth　锯齿形的
scale [skeil] *n*. 比例;缩尺;比例尺;等级;样卡
scarf [skaːf] *n*. 围巾,披巾,披肩,头巾,领巾,腰巾
scheme [skiːm] *n*. 计划,方案
scorching [ˈskɔːtʃiŋ] *n*. 烫黄,烫焦
Scottish kilt　苏格兰褶裥短裙
scratchy [ˈskrætʃiː] *adj*. 使人发痒的,扎人的
seal [siːl] *v*. 封印,密封,隔离;图记;记号
sealer [ˈsiːlə] *n*. 保护层
seam allowances　缝接允差,缝合允许量,缝头,缝份
seam ripper　拆线器
seam roll　袖馒头,袖烫垫
seam slippage　跳线
second-hand　用旧的,二手的
sedentary [ˈsedəntəri] *adj*. 少动的;固定于一点的
seductive [siˈdʌktiv] *adj*. 诱惑的,引人注意的,有魅力的
seep through　渗透,渗过
self [self] *n*. 自己,自我;本性
self-certification　行认定
self-trimming　同料边饰
selvedge [ˈselvidʒ] *n*. 织边,布边,边缘
semi-fitting　半紧身的
semi-precious gemstone　半宝石的
sensitive [ˈsensitiv] *adj*. 敏感的,易受伤害的
sensor [ˈsensə] *n*. 传感器
sequentially [sɪˈkwenʃəli] *adv*. 持续地,连续地
set-in　另外缝上的,装袖
shade [ʃeid] *n*. 色泽,色光,色调,色度,明暗程度
shape retention　形状保持性
shape [ʃeip] *n*. 形状,外形;(人体的)特有形状;体形,身段
shaped [ʃeipt] *adj*. 成型的
sheath [ʃiːθ] *n*. 鞘,套,外层覆盖(物),外包物
sheer [ʃiə] *n*. 透明薄织物;透明薄纱
shell fabric　面料
shell [ʃel] *n*. 贝壳,壳
shift [ʃift] *v*. 变化,更替,转移,改变　*n*. 开关;变化
shiny [ˈʃaini] *adj*. 发光的;光亮的;有光泽的
shirring [ˈʃəːriŋ] *n*. 多层收皱,抽褶,平行皱缝
SHLDR Width　肩宽
SHLDR.　shoulder 的缩写
shoe [ʃuː] *n*. 鞋子
shoulder dart　肩省
shoulder neck point　(SNP)侧颈点
shoulder pads　垫肩,肩衬,护肩
shoulder point　(SP)肩点
shoulder [ˈʃəuldə] *n*. 肩宽,肩,肩胛骨
showroom [ˈʃəurum] *n*. (样品)陈列室,展览室
shrink [ʃriŋk] *v*. 收缩,缩水
shuttle [ˈʃʌtl] *n*. (缝纫机的)滑梭
side seam　(SS)侧缝线
signal [ˈsignəl] *n*. 信号,暗号,标志
significant [sigˈnifikənt] *adj*. 有意义的,重要的,重大的
Silesia [saiˈliːzjə] *n*. 西里西亚里子布
silhouette [ˌsiluːˈet] *n*. 轮廓

silk screen printing 丝网印
simplistic [sim'plistik] *adj*. 简单化的
simultaneous [ˌsiməl'teinjəs] *adj*. 同时发生的,同步的
singe [sindʒ] *v*. 烧去(布匹的)茸毛
single-ply 单层
size grade chart 尺码表
sizeable ['saizəbl] *adj*. 相当大的,可观的
sketch [sketʃ] *n*. 草图,速写
skin [skin] *n*. 毛皮,兽皮
skinny ['skiniː] *adj*. 消瘦的;细窄的
skipants [skiːpænts] *n*. 滑雪裤
skirt marker 裁裙片样板
sku number 单品数量
skullcap ['skʌlkæp] *n*. 无檐便帽,瓜皮帽
sleeve board 压袖板,烫袖板,袖子烫板;烫马
sleeve central line (SCL)袖中线
sleeve head 袖山头
sleeve opening 袖口
sleeve sewing notch 袖片对位点(刀眼)
slender ['slendə] *adj*. 细瘦的,苗条的
slide tab 拉链头子,拉链滑块
slight [slait] *adj*. 细小的,轻微的
slightly ['slaitli] *adv*. 些微地,苗条地,瘦长地
slim [slim] *adj*. 细长的,苗条的,纤细的
slip [slip] *n*. 活络里子,套裙,女式长衬裙,儿童围兜
slippery ['slipəri] *adj*. 易脱落的,手感滑的
slit [slit] *n*. 缝,裂口,槽; *v*. 开衩
sloper ['sləupə] *n*. 服装尺寸样板
smart card 智能卡
smart [smaːt] *adj*. 漂亮的
smooth [smuːð] *adj*. 光滑的,滑爽的
smooth-surfaced 光滑表面的
snap [snæp] *n*. 按钮,钩扣
snugly ['snʌgli] *adv*. 舒适地,整洁干净地,紧密地
solely ['səʊli] *adv*. 独自地,单独地
solvent ['sɔlvənt] *n*. 溶剂,溶媒
sound-wave 声波
spaced [speis] *adj*. 间隔的
spaghetti strap 细肩带
spandex ['spændeks] *n*. 弹力纤维
specialize in 专门从事,专门研究
specialty store 专卖店
specialty ['speʃəlti] *n*. 专长,专业,特点
specification [ˌspesifi'keiʃən] *n*. 规格,规范;明细表;(产品)说明书
spectacular [spek'tækjulə] *n*. 公开展示的,展览物
spinneret ['spinəret] *n*. 吐丝器,喷丝头
spinning ['spiniŋ] *n*. 纺纱
splash out 挥霍钱财
sponge rubber 橡胶海面,多孔橡胶,泡沫橡胶
spongy ['spʌndʒiː] *adj*. 海绵状的
spool [spuːl] *n*. 线团
sportswear ['spɔːtswɛə] *n*. 便装;运动服装
spot 污迹;(图案)点子,斑点
spring [spriŋ] *v*. 弹回,反弹
squeeze [skwiːz] *v*. 挤,塞,压进,挤入
squint [skwint] *v*. 迷眼;斜眼
stability [stə'biliti] *n*. 稳定性
stack [stæk] *v*. 叠,堆,堆积于,把……堆积于
stain and spot resistance 防沾污性
stance [stæns] *n*. 姿态,态度,立场
standardization [ˌstændədai'zeiʃən] *n*. 使合标准;使标准化
standing ['stændiŋ] *n*. 地位,身份,名声
staple ['steipl] *n*. 主要产品,原材料
starch [staːtʃ] *v*. 浆洗
startup ['staːtˌʌp] *n*. 启动
state-of-the-art 达到最新技术发展水平的
statute ['stætjuːt] *n*. 法令,法规
stay tape 定位带,胸衬条,牵条,过桥布
stay [stei] *n*. 滚边,窄带

steam and dry iron　蒸汽电熨斗
steam cleaning　蒸汽喷洗；蒸汽清洗
steer [stiə] *v*. 引导，控制
sticker ['stikə] *n*. 缝纫机；缝纫工
stiff hand　硬手感
stiffness ['stifnis] *n*. 硬挺性，硬挺度
stipulate ['stipjəˌleit] *v*. 确定，保证，规定；约定
stitch [stitʃ] *n*. 针脚，线迹；缝法，缝线
stocking ['stɔkiŋ] *n*. 长袜
straight knife　直条裁剪刀
straight [streit] *adj*. 直线的；直筒的
straight-stitch　直线线迹
strapless ['stræplis] *adj*. 无吊带，无肩带
strategy ['strætidʒi] *n*. 战略，策略
stratum ['straːtəm] *n*. 阶层
street fashion　街头服装，街头服饰
strength [streŋθ] *n*. 强度
stretch [stretʃ] *n*. 弹力，弹性，松紧
stretchability [ˌstretʃə'biliti] *n*. 拉伸性；延伸性；拉伸性
stretchable [stretʃəbl] *adj*. 有弹性的
striking ['straikiŋ] *adj*. 引人注目的，吸引人的
string [striŋ] *n*. 细绳，绳索
stripe [straip] *n*. 条子，条纹
strive to　努力，致力于……
style No.　款式号码
style [stail] *n*. 型，款式，式样；时尚；类型；气派，风度，格调
style-line　造型线，款式结构线
stylish ['stailiʃ] *adj*. 时尚的，时髦的
stylized sketch　效果图
subculture ['sʌbˌkʌltʃə] *n*. 亚文化
subdue [səb'djuː] *v*. 使屈服，压制，克制；降低
subscription [səb'skripʃən] *n*. 订阅
substitute ['sʌbstitjuːt] *n*. 代用品；代替，替代；*adj*. 代用品的
substitution [ˌsʌbsti'tjuːʃən] *n*. 代替，替换，代替物
subtly ['sʌtli] *adv*. 柔和地，巧妙地，精细地
suction ['sʌkʃən] *n*. 吸入，抽吸，抽气通风
suede [sweid] *n*. 人造麂皮，小山羊皮，起毛皮革，绒面革
suitability [ˌsjuːtə'biliti] *n*. 合适，适当，相配，适宜性
suite [swiːt] *n*. （软件的）套件
sumptuous ['sʌmptʃuːəs] *adj*. 豪华的，奢侈的
sunglass ['sʌnˌglæs] *n*. 太阳镜
supermodel ['sjuːpəˌmɔdəl] *n*. 超级模特
supersede [ˌsjuːpə'siːd] *v*. 代替，接替，更替
supplement ['sʌplimənt] *n*. 增补（物），补充（物）
supplementary [ˌsʌpli'mentəri] *adj*. 补充的
supplier [sə'plaiə] *n*. 厂商，供应商
supply chain　供应链
suppress [sə'pres] *v*. 削弱，压制
surrealist-inspired　超现实主义灵感来源的
suspenders [sə'spendəz] *n*. 背带，吊带
swatch [swɔtʃ] *n*. 小块样布；样品，样本
sweater ['swetə] *n*. 运动衫；针织套衫；毛衣，毛线衫
sweep [swiːp] *n*. 下摆围
swimsuit ['swimˌsuːt] *n*. 女游泳衣
swing [swiŋ] *n*. 曲线形轮廓
symbol ['simbəl] *n*. 象征，表征
synonymous [si'nɔnəməs] *adj*. 同义的
synthesize ['sinθiˌsaiz] *v*. 综合；合成
synthetic fiber　合成纤维

T

tack [tæk] *n*. 粗缝，假缝　*v*. 假缝；系、扎、扣、拴

tactile [ˈtæktəl] *n.* 手感，触感
tag [tæg] *v.* 标上……标签，标签
tailor's chalk （裁缝用）划粉
tailor's ham 布馒头，馒头烫垫
tailored suit 西式套装，精做西装
tailored [ˈteiləd] *adj.* 定做的，合身的，精做的，精致的
tailoring [ˈteiləriŋ] *n.* 缝制，裁剪业；裁缝工艺，成衣工艺
take advantage of 利用
take for granted 认为……理所当然
take into account 把……考虑进去
tape measure 软尺，带尺，卷尺
tape [teip] *n.* 狭幅织物，狭带，卷尺 *v.* 贴边，镶边，牵条
taper [ˈteipə] *adj.* 锥形的；使成锥形
tassel [ˈtæsəl] *n.* 穗，流苏
taste [teist] *n.* 式样，风格，审美力，欣赏力
technologist [tekˈnɔlədʒist] *n.* 工艺师
tedious [ˈtiːdiəs] *adj.* 单调乏味的，令人生厌的，繁重的
teenage wear 少年装
teleconference [ˈtelikɔnfərəns] *n.* （通过电话，电视的）电信会议
template [ˈtemplit] *n.* 模板
temporary [ˈtempərəri] *adj.* 暂时的
tempt away 诱惑，引诱
tensile strength 抗张强度
tension [ˈtenʃən] *n.* 拉力，张力
terry cloth *n.* （毛巾，围巾等）两端留有绒穗的物品
test fit 裁剪前的试样
textile [ˈtekstail] *n.* 纺织品，织物，纺织的，纺织原料
texture [ˈtekstʃə] *n.* 织物组织；织物质地，纹理，肌理
the Arctic Circle 北极圈
the failure rate 失败率
the lock stitch machine 锁式线迹缝纫机
the over-lock machine 包缝机，拷边机
the safety stitch over-lock machine 安全线迹包缝机
the zigzag machine 锯齿形锁缝缝纫机，曲折缝缝纫机
theatrical [θiːˈætrikəl] *adj.* 夸张的，戏剧性的
theoretically [θiəˈretikli] *adv.* 理论上，理论地
thermal retention 保暖性
thermoplastic [ˌθəːməˈplæstik] *adj.* 热塑性的 *n.* 热塑性塑料
thick and thin 粗细交替的
thick [θik] *adj.* 厚，浓，深，密
thickness [ˈθiknis] *n.* 厚度，密度
thimble [ˈθimbl] *n.* （手工缝纫用的）顶针，针箍
thread clippers 纱剪刀
thread loop 线环
thread [θred] *n.* 线
tie [tai] *v.* 系，打结，扎，绑，捆
tier [tiə] *n.* 等级
tights [taits] *n.* 裤袜，紧身衣裤
time-base 时间基线；时间坐标
timeline 时间轴
tint [tint] *n.* 色，色彩，色泽，色度；淡色
tissue paper 薄纸
tissue [ˈtisjuː] *n.* 薄纱，餐巾纸
toggle [ˈtɔgl] *n.* 套环，套索扣
toile pattern 样衣纸样
toile [twaːl] *n.* 试穿服装，样衣
TOL. tolerance [ˈtɔlərəns] 的缩写 *n.* 公差；容限；容许数；宽松度
top & bottom （织物）上下端
top hip line （THL）上臀线
tops [tɔps] *n.* 上装
tortoiseshell [ˈtɔːtəʃel] *n.* 龟甲，玳瑁
tracing paper 描图纸
tracing wheel 插盘；点线轮，描样手轮

trademark [ˈtreidmaːk] *n*. 商标，商标，牌号

transformative [trænsˈfɔːmətiv] *adj*. 使变化的，有变形力的

tremendous [triˈmendəs] *adj*. 巨大的

trend setter （服装式样）创新人

trend [trend] *n*. 时尚，流行，趋势

triangular [traiˈæŋgjulə] *adj*. 三角形的

tricot knits 经编织物

trim to 修饰使适合……

trousers [ˈtraʊzəz] *n*. 裤子，长裤，西装裤

truing [ˈtruːiŋ] *n*. 描实

trunk show 展销

try out 试验，试穿

T-shirt [ˈtiː ˌʃəːt] *n*. 体恤衫，短袖圆领汗衫

tubular [ˈtuːbjələ] *adj*. 管状的

tuck [tʌk] *n*. 横裥，活褶，塔克 *v*. 打褶，打裥

turban [ˈtɜːbən] *n*. 女式头巾，穆斯林头巾

tweed [twiːd] *n*. 粗花呢

twill [twil] *n*. 斜纹布，斜纹图案

twine [twain] *v*. 合股，搓，交织，缠绕

twist [twist] *v*. 捻，拧，编织

two-dimensional 二维的，平面的

Tyvek [ˈtaivik] *n*. 高密度聚乙烯合成纸

U

ultra-sonic 超音波

uncut [ˈʌnˈkʌt] *adj*. 不裁剪的，未经裁剪的

under pressing 缝前熨烫，半成品熨烫

underarm point 腋点

underarm [ˈʌndəraːm] *n*. 腋下

under-buy 低价购买

undercollar [ˌʌndəˈkɔlə] *n*. 领里，领下片

underestimate [ˌʌndərˈestimeit] *v*. 低估，估计不足

underlap [ˌʌndəˈlæp] *n*. 里襟

underlining [ˌʌndəˈlainiŋ] *n*. 衬里

undermine [ʌndəˈmain] *v*. 在……下方

underneath [ˌʌndəˈniːθ] *v*. 在下面，在……的下面

under-purchase 低价购置

understitch [ˌʌndəˈstitʃ] *v*. 暗缝

underwear [ˈʌndəwɛə] *n*. 内衣，衬衣

uneven [ʌnˈiːvən] *adj*. 不均匀的

unity [ˈjuːniti] *n*. 和谐，协调，统一

unloading [ˈʌnˈləudiŋ] *n*. 卸载

unwind [ʌnˈwaind] *adj*. 松散的，未卷绕的

unwrinkle [ˌʌnˈriŋkl] *v*. 将（皱纹）弄平

up and down 上上下下的

update [ʌpˈdeit] *v*. 修正，更新

upholstery [ʌpˈhəʊlstəriː] *n*. 室[车]内装饰（品）

uppermost [ˈʌpəˌməʊst] *adj*. 最上面的，最高的

utilization [ˌjuːtilaiˈzeiʃən] *n*. 利用，使用，应用

V

vacuum-com-pressed 吸风烫台

value [ˈvælju] *n*. 明度

value-led 价值引导的

variable [ˈvɛəriəbl] *n*. 变数，变化因素

velvet [ˈvelvit] *n*. 经绒，丝绒，立绒，天鹅绒

velveteen [ˈvelviˈtiːn] *n*. 纬绒，棉绒，平绒

vendor [ˈvendə] *n*. 厂商，卖家

vertex [ˈvɜːˌteks] *n*. 顶点

vertical [ˈvəːtikəl] *adj*. 垂直的，竖式的，直立的，纵的

via [ˈvaiə] *prep*. 通过

videoconferencing [ˌvidiəuˈkɔnfərənsiŋ] *n*. 视频会议

vinyl [ˈvainil] *n*. 聚乙烯基织物

virtual [ˈvəːtjuəl] *adj*. 实际上的,实质上的
visual illusion　视错觉,错视
visual merchandising　展示销售
V-neck　V 形领
volatile [ˈvɔlətail] *adj*. 可变的,不稳定的
volume [ˈvɔljuːm] *adj*. 大量的,大批量
voluntary standard　自愿执行标准
voluntary [ˈvɔləntəri] *adj*. 自发的,自愿的

W

wadding [ˈwɔdiŋ] *n*. 填絮,软填料,衬垫
waist dart　腰省
waist line　(WL)腰线
waist [weist] *n*. 腰围
waistband [ˈweistbænd] *n*. 腰带
waistline [ˈweistˌlain] *n*. 腰节;腰围线,腰节线
wardrobe [ˈwɔːdrəub] *n*. (个人的)全部服装,行头;衣柜,衣橱
warehouse [ˈweəhaus] *n*. 仓库
warp [wɔːp] *n*. 经向,经纱
warp-knitted fabric　经编针织物
warranty [ˈwɔrənti] *n*. 保证(书),根据理由,授权
washability [wɔʃəˈbiliti] *n*. 可洗性
washing [ˈwɔʃiŋ] *n*. 洗涤
waste [weist] *n*. 废料,废纱
watch out for　密切注意,戒备,提防
watch [wɔtʃ] *n*. 手表
water repellency　拒水性
waterproofing [ˈwɔːtəpruːfiŋ] *n*. 防水,绝湿
wax [wæks] *n*. 蜡
wear [weə] *n*. 服装,衣服,穿戴物
weatherproof [ˈweðəpruːf] *adj*. 防风雨的
weather-wear　风雨衣
weave [wiːv] *n*. 梭织,织物,织法,编织式样　*v*. 编织
wedge [wedʒ] *n*. 楔形
wedge-shaped　楔形
weed out　清除
weft [weft] *n*. 纬向,纬纱
weft-knitted fabric　纬编针织物
well-known　众所周知的,有名的
well-ventilated　透气的
whilst [wailst] *conj*. 当……时候
white dye　增白,加白
wholesaler [ˈhəulseilə] *n*. 批发商
wicking ability　吸附能力
width [widθ] *n*. 幅宽
window display　橱窗展示,橱窗陈列
women-wear　妇女服装,女式服装
wood pulp　木纸浆
working-class　工薪族
workmanship [ˈwəːkmənʃip] *n*. 手工,工艺
wrap [ræp] *v*. 包裹,围裹
wrinkle resistancy　抗皱性,抗皱处理
wrinkle [ˈriŋkl] *n*. 皱褶,折皱
wrinkle-free　不皱的
wrinkle-resistant　防皱处理,耐褶皱性
wrist [rist] *n*. 腕,腕部

Y

yard stick　码尺,直尺
yarn [jaːn] *n*. 纱,纱线
yoke [jəuk] *n*. 过肩,覆肩,育克,约克
Y-shaped　Y 形的

Z

zig-zag　曲折线条,锯齿形,曲折形,人字形
zipper [ˈzipə] *n*. 拉链
zoom in　放大
zoom out　缩小

附 录

一、服装各部位名称

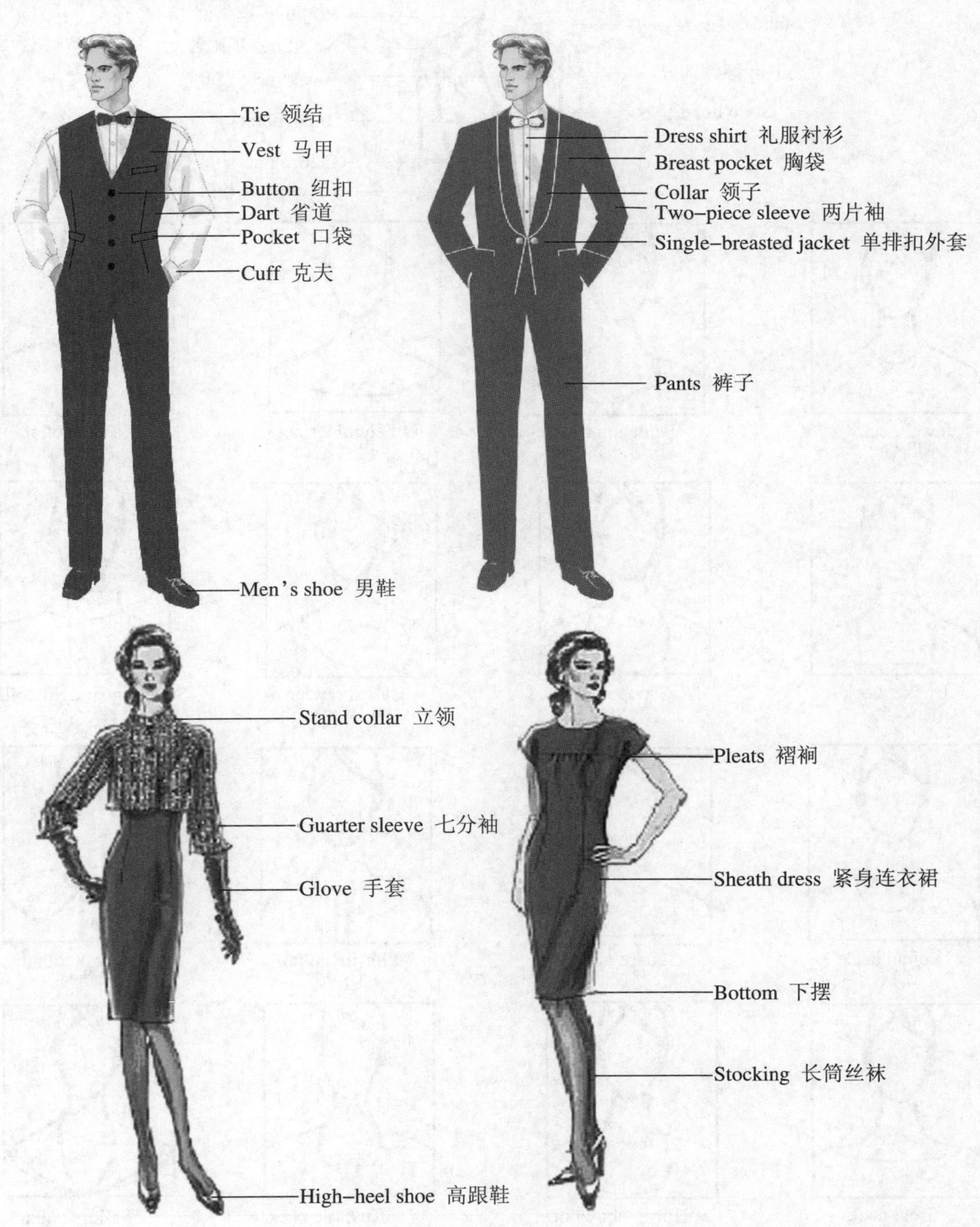
Tie 领结
Vest 马甲
Button 纽扣
Dart 省道
Pocket 口袋
Cuff 克夫
Men’s shoe 男鞋
Dress shirt 礼服衬衫
Breast pocket 胸袋
Collar 领子
Two-piece sleeve 两片袖
Single-breasted jacket 单排扣外套
Pants 裤子
Stand collar 立领
Guarter sleeve 七分袖
Glove 手套
High-heel shoe 高跟鞋
Pleats 褶裥
Sheath dress 紧身连衣裙
Bottom 下摆
Stocking 长筒丝袜

二、服装领子的名称与种类

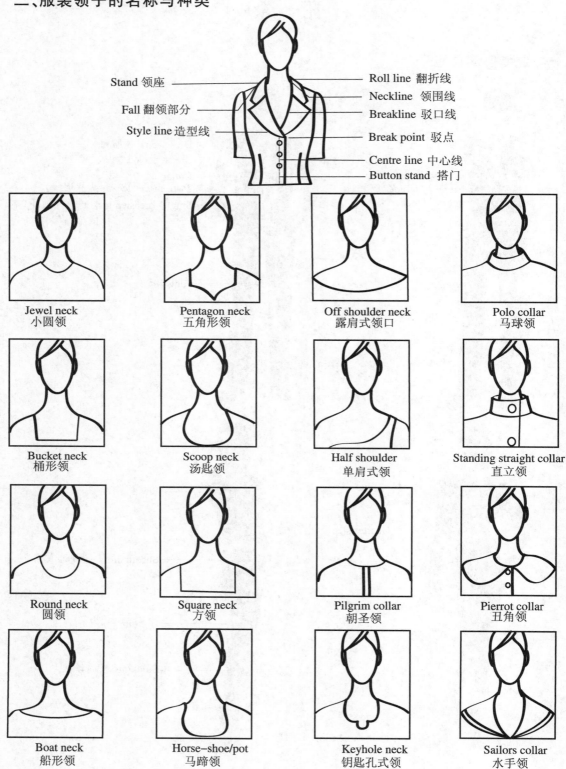

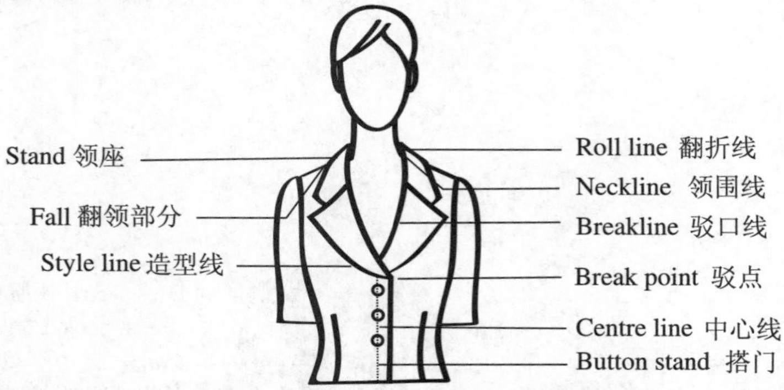

Stand 领座
Fall 翻领部分
Style line 造型线
Roll line 翻折线
Neckline 领围线
Breakline 驳口线
Break point 驳点
Centre line 中心线
Button stand 搭门

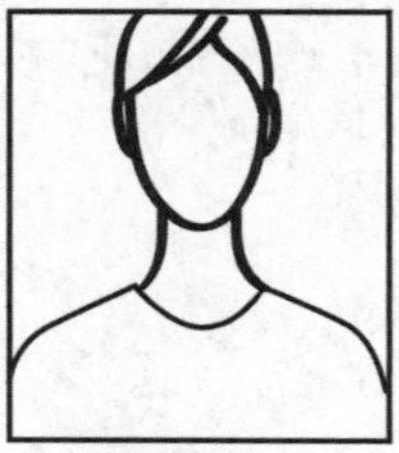

Jewel neck
小圆领

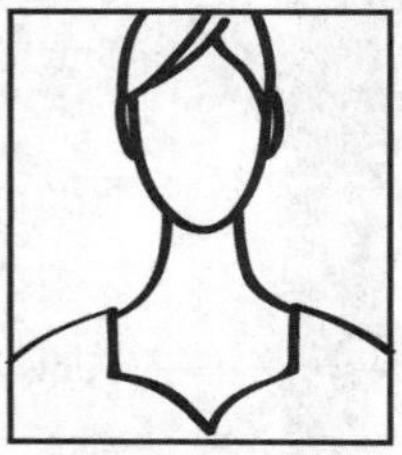

Pentagon neck
五角形领

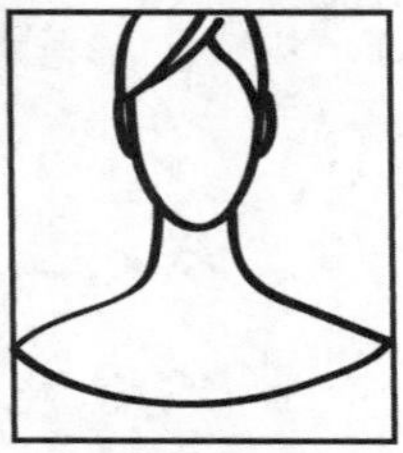

Off shoulder neck
露肩式领口

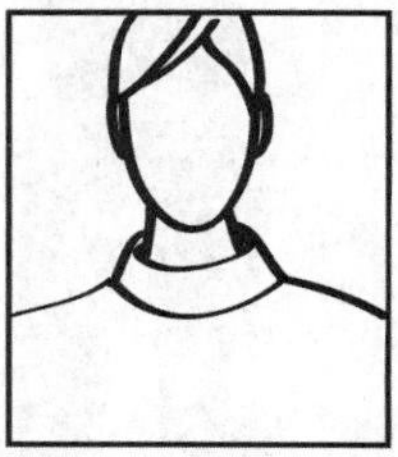

Polo collar
马球领

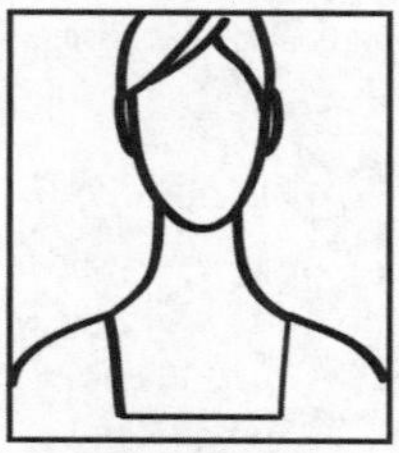

Bucket neck
桶形领

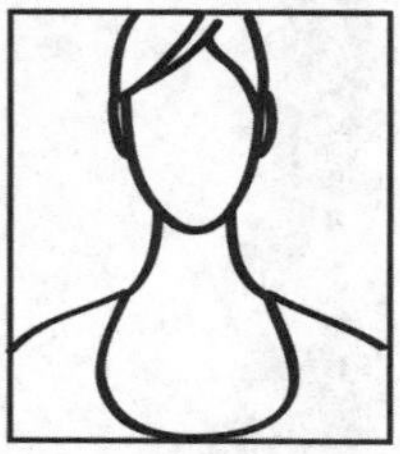

Scoop neck
汤匙领

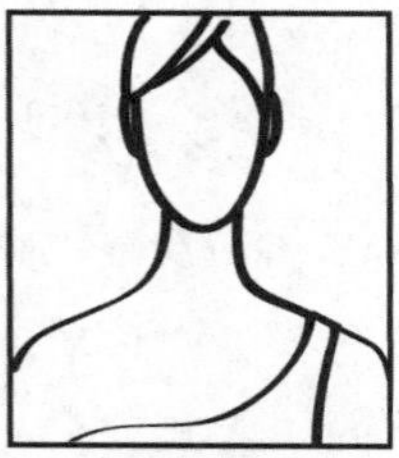

Half shoulder
单肩式领

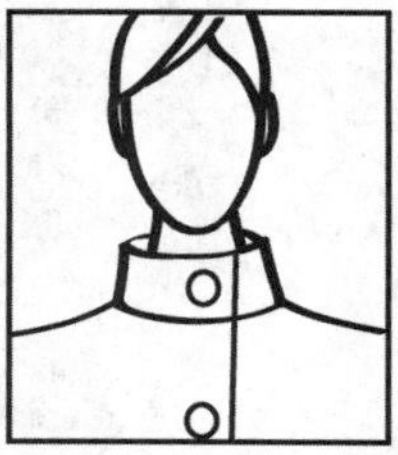

Standing straight collar
直立领

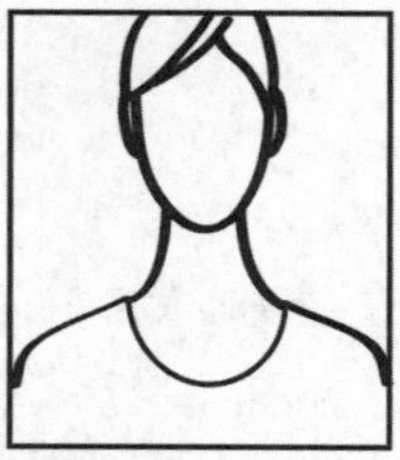

Round neck
圆领

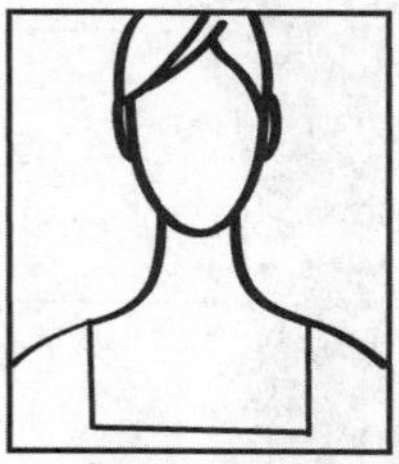

Square neck
方领

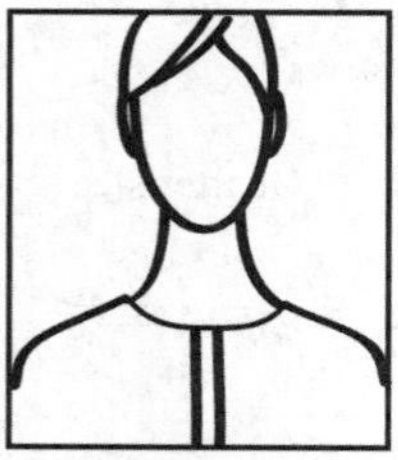

Pilgrim collar
朝圣领

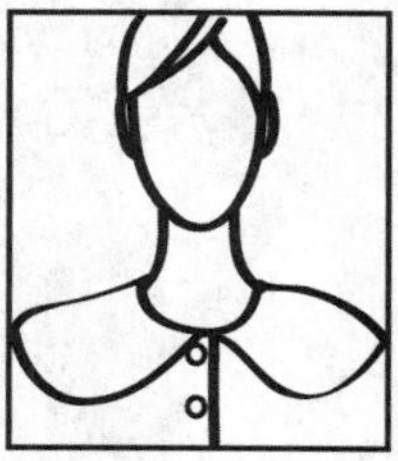

Pierrot collar
丑角领

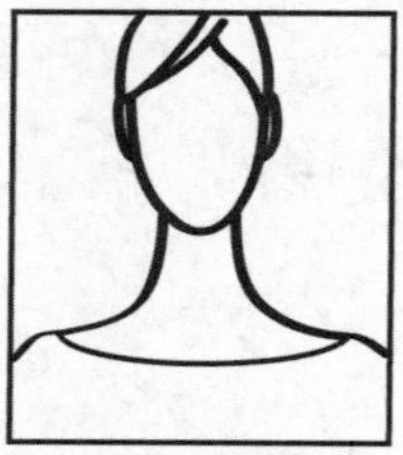

Boat neck
船形领

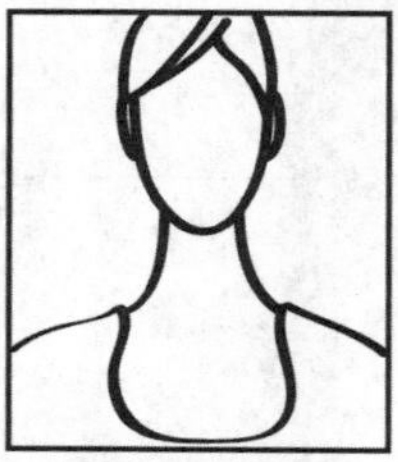

Horse-shoe/pot
马蹄领

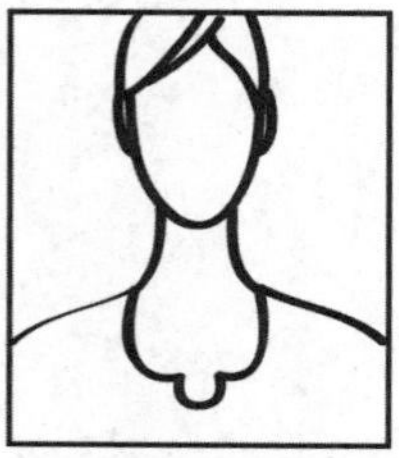

Keyhole neck
钥匙孔式领

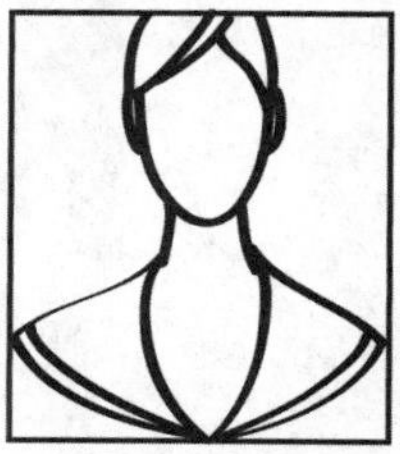

Sailors collar
水手领

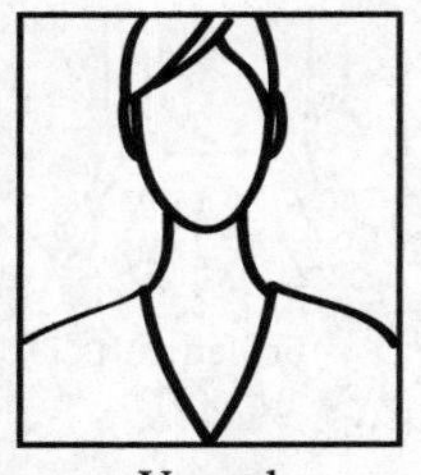
V–neck
V 字领

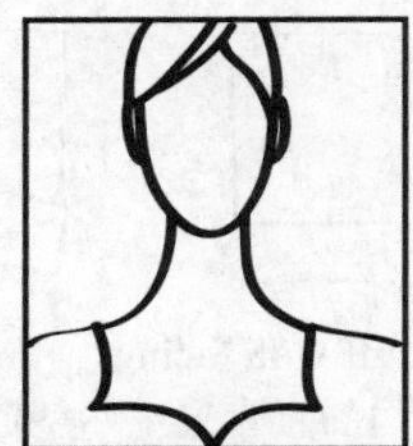
Sweet heart
鸡心领

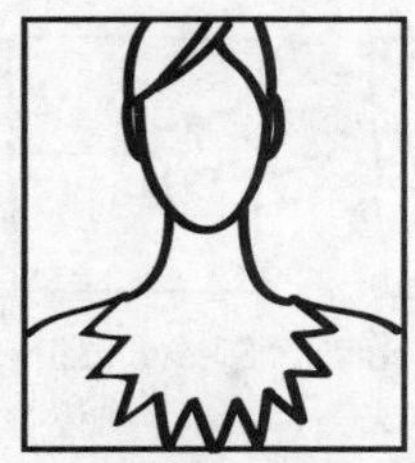
Zigzag neck
锯齿领

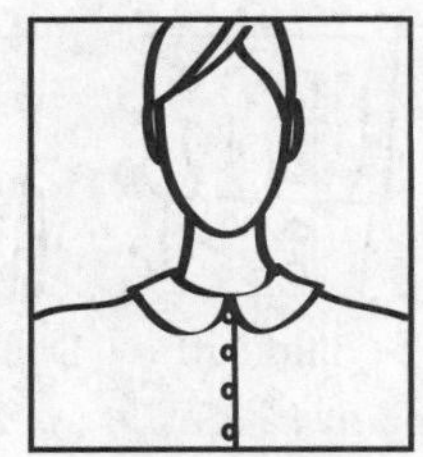
Flat collar
平领

Scallop neck
荷叶边领

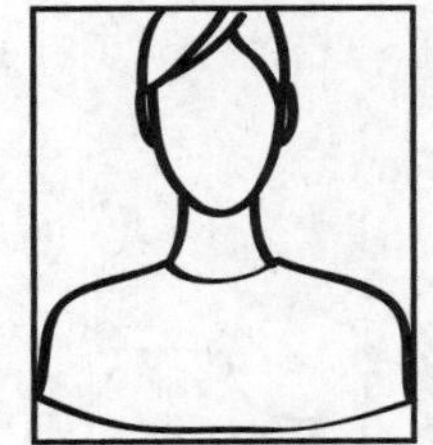
Bertha neck
宽圆领

Convertible collar
两用领

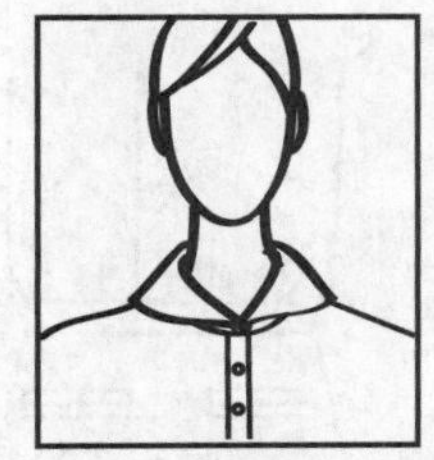
Wing collar
翼领

三、袖子与克夫的种类

Inset cap sleeve
袖山装袖

Puff sleeve
泡泡袖

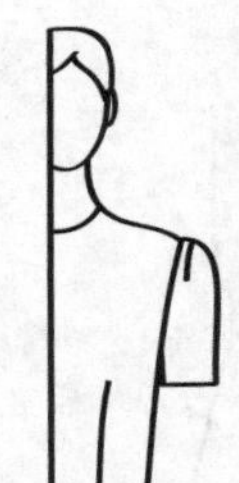
Strapped head sleeve
窄条袖

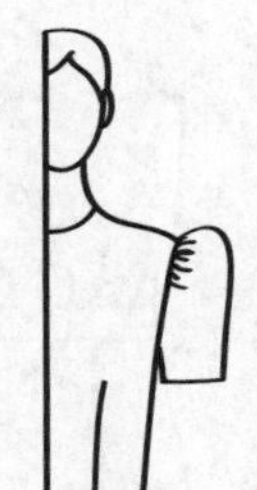
Gathered head sleeve
有褶袖山袖

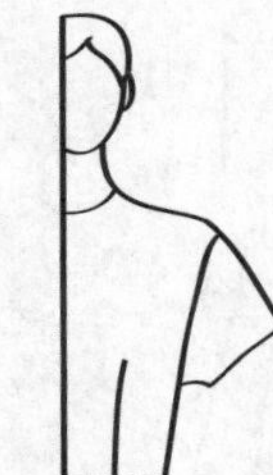
Dolman sleeve
德尔曼袖

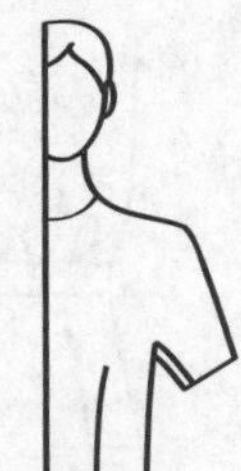
Cap sleeve with gusset
带袖裆的袖山袖

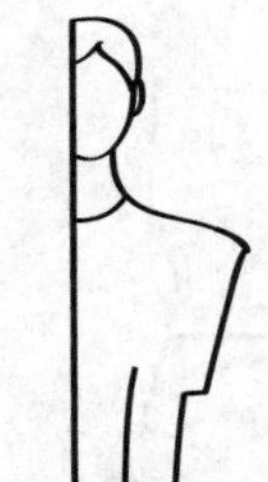
Cap sleeve
袖山袖

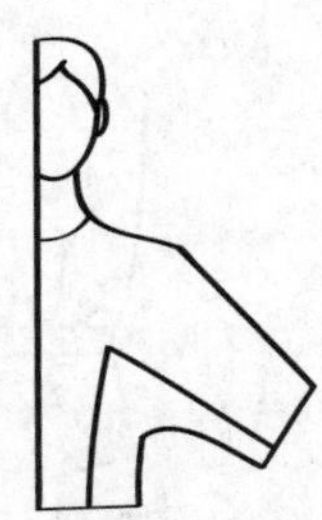
Kimono style with yoke
有育克的连袖

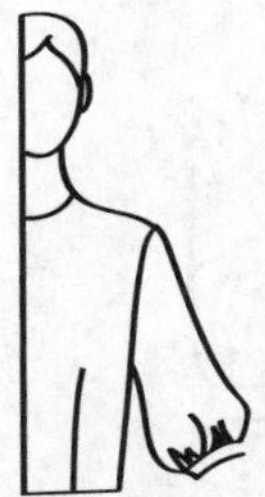
Gathered into cuff
有褶克夫袖

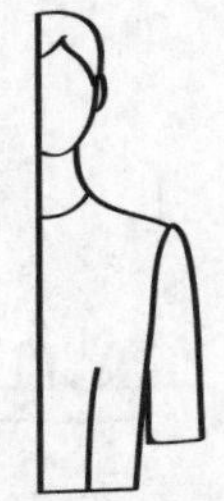
Fitted short sleeve
合体短袖

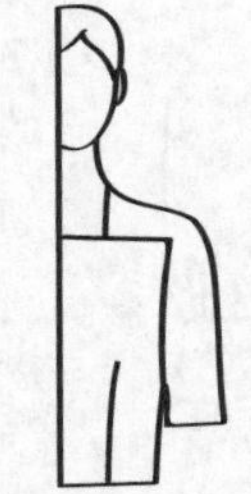
Extended sleeve with yoke
带育克的过肩袖

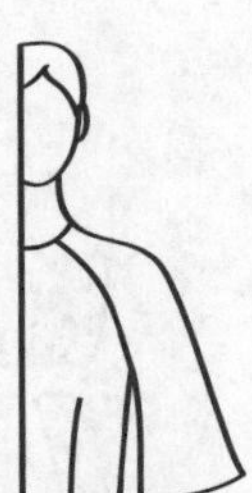
Flared raglan sleeve
喇叭插肩袖

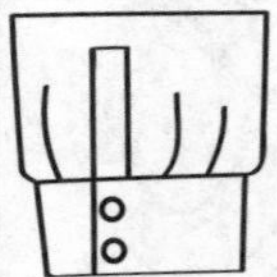
Shirt cuff
衬衫克夫

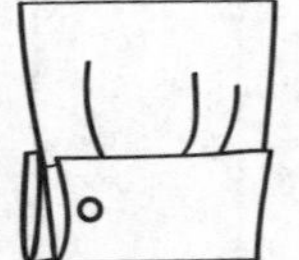
Double shirt cuff
折叠衬衫克夫

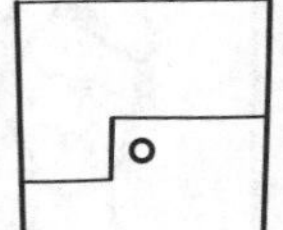
Sleeve facing
袖子贴边

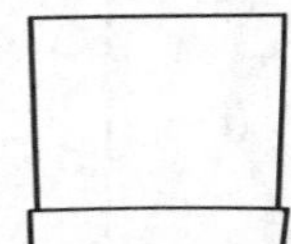
Straight cuff with facing
有贴边的直克夫

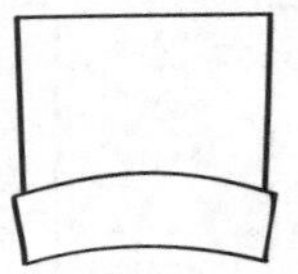
Shaped cuff
变形克夫

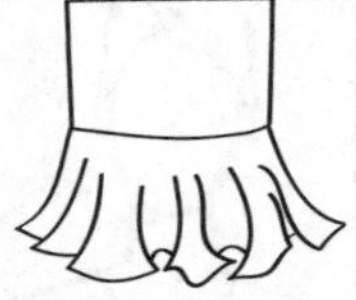
Frilled cuff
褶边克夫

四、口袋的种类

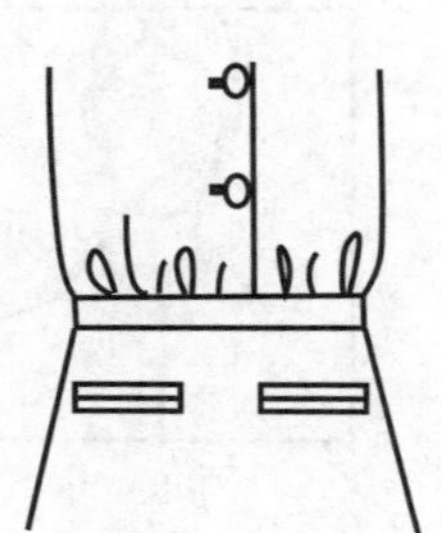
Welt pocket
开线袋

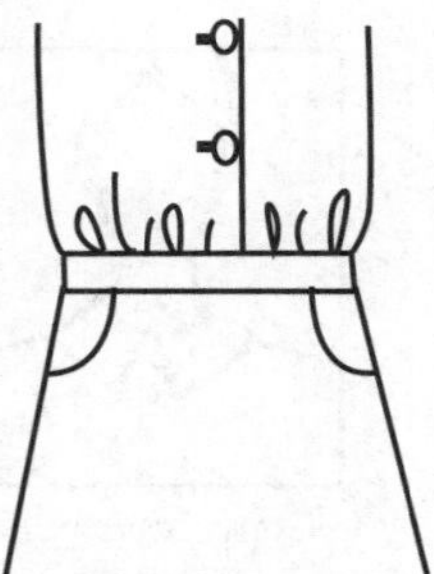
Western pocket
嵌线袋

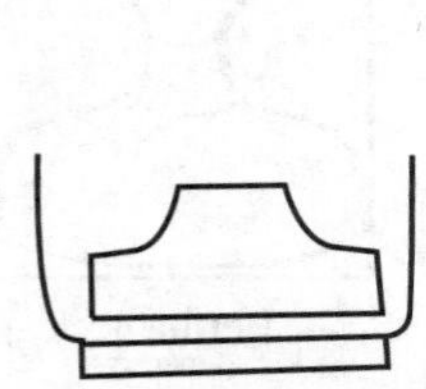
Kangaroo pocket
袋鼠式口袋

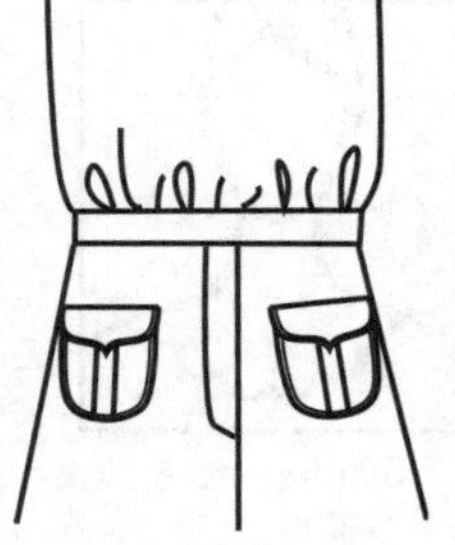
Safari pocket
猎装口袋

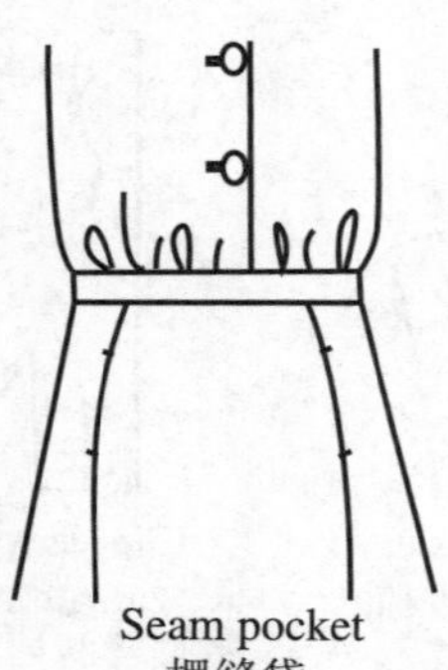
Seam pocket
摆缝袋

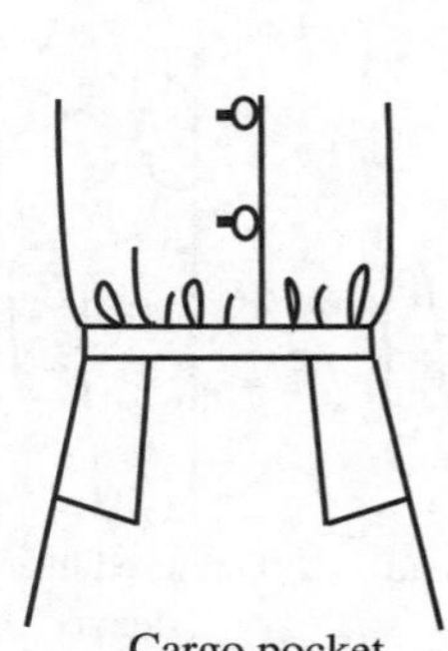
Cargo pocket
大贴袋

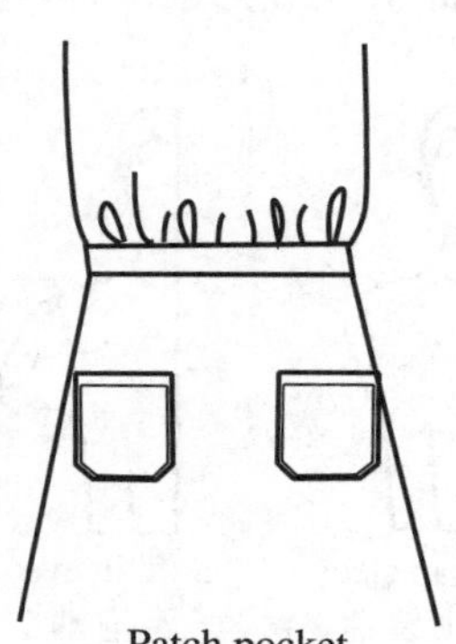
Patch pocket
贴袋

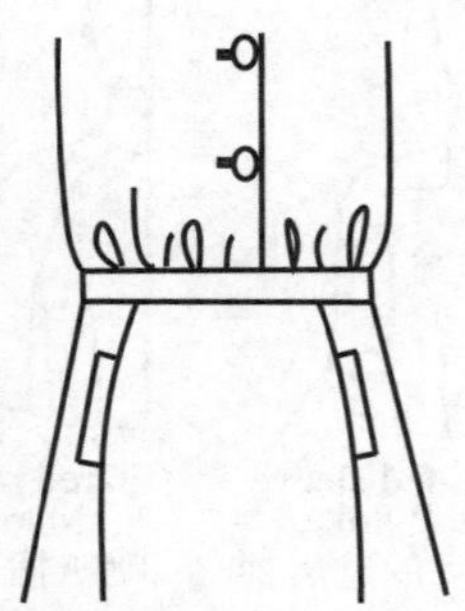
Flap pocket
盖式口袋

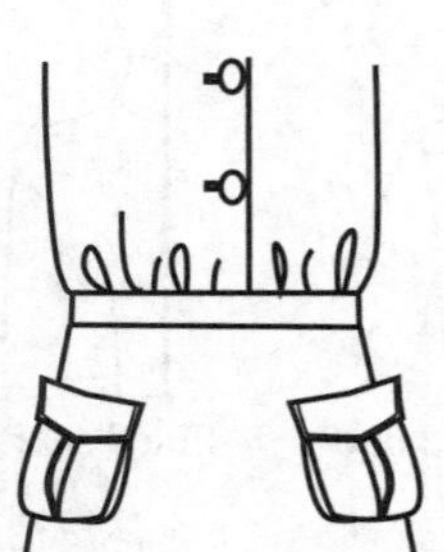
Bellows pocket
风箱式口袋

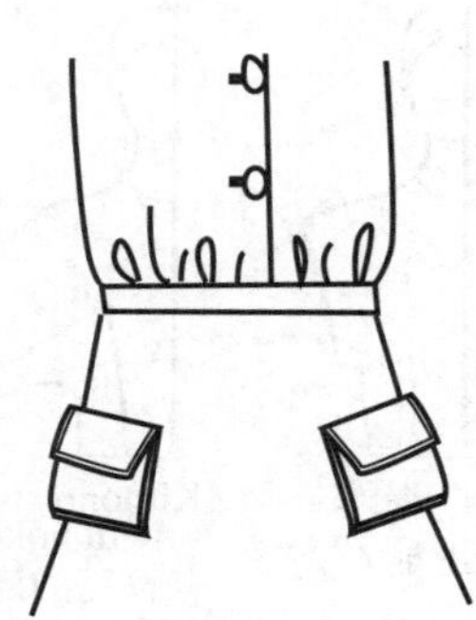
Envelop pocket
信封式口袋

五、裙子的种类

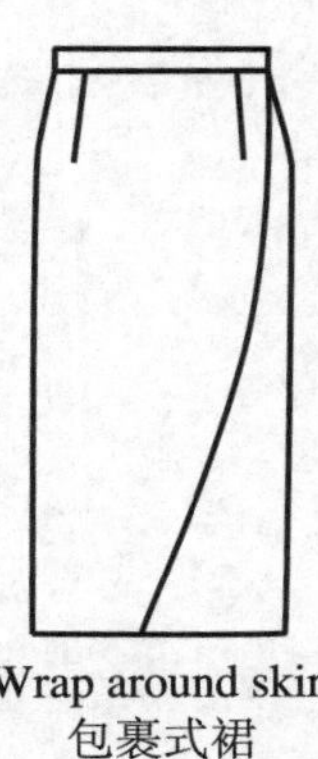
Wrap around skirt
包裹式裙

Tulip/Petal skirt
郁金香式/花瓣式裙

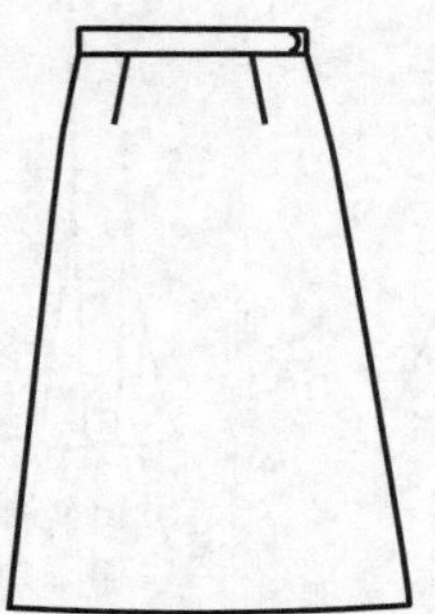
Aline skirt
A字裙

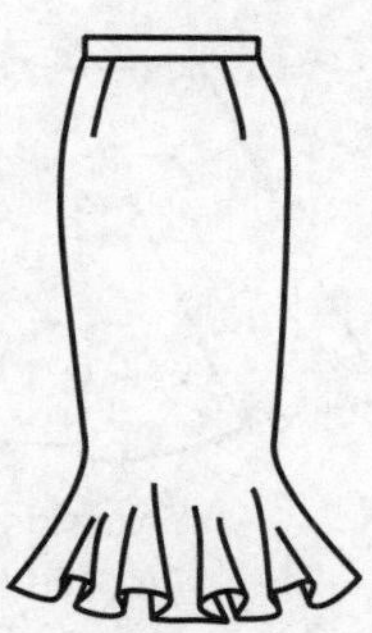
Trumpet skirt
喇叭裙

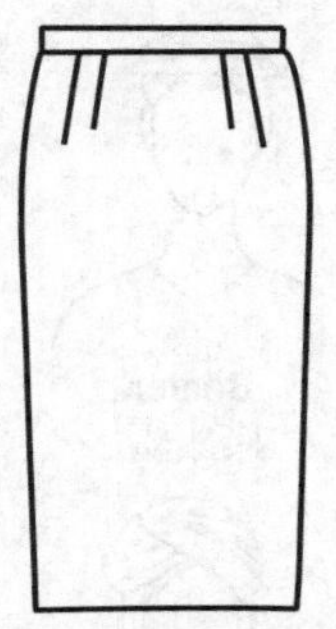
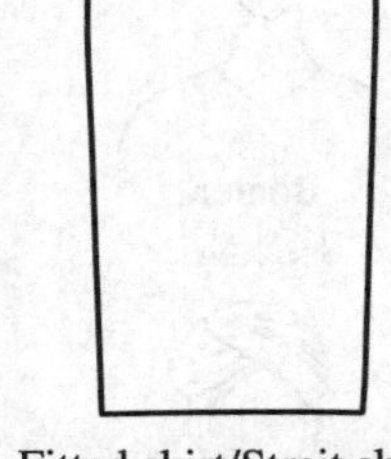
Fitted skirt/Strait skirt
直筒裙

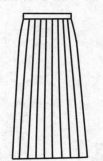

Skirt with allround pleats
百褶裙

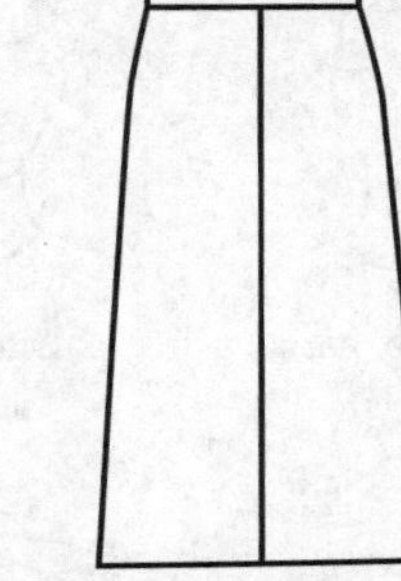

Four gored skirt
四片裙

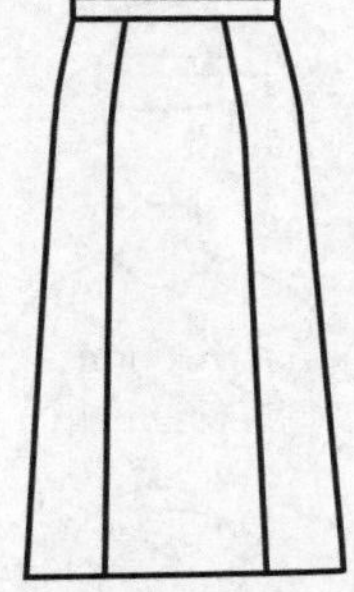
Six gored skirt
六片裙

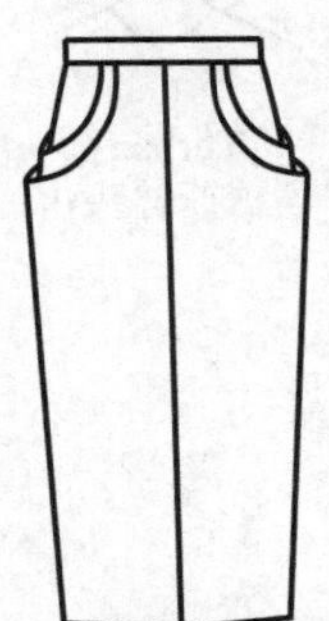

Cowl skirt
荡褶裙

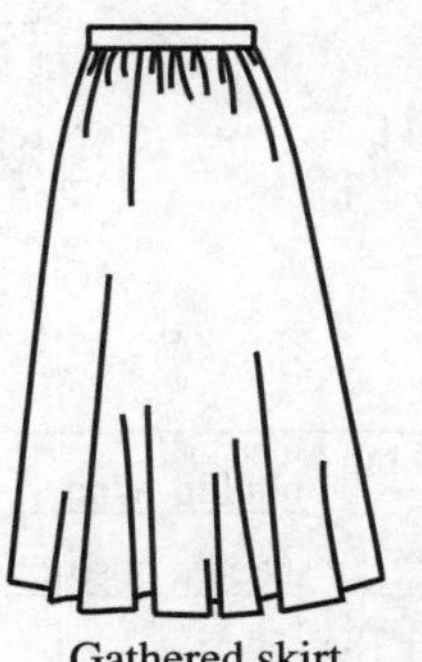
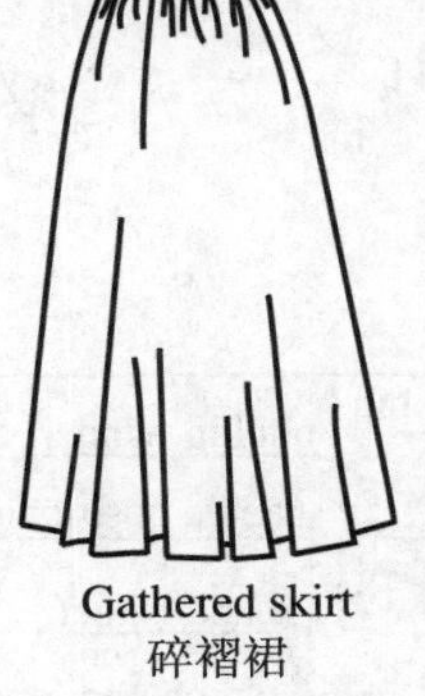
Gathered skirt
碎褶裙

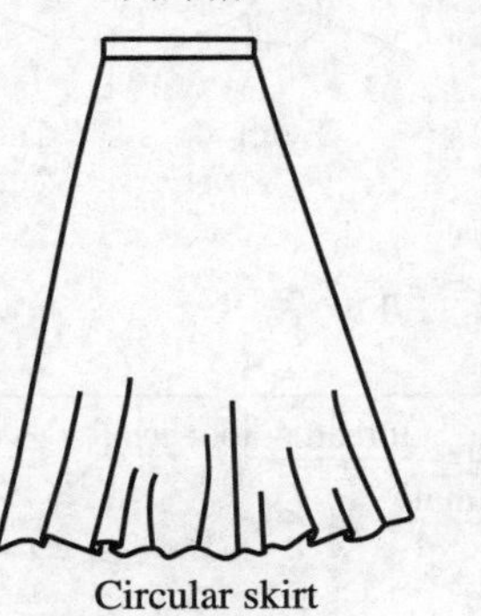
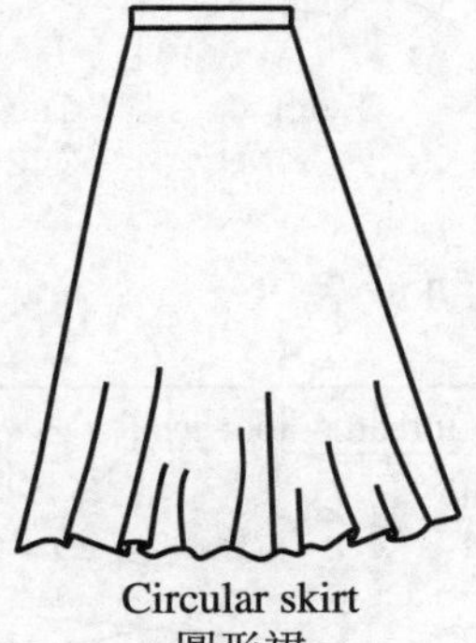
Circular skirt
圆形裙

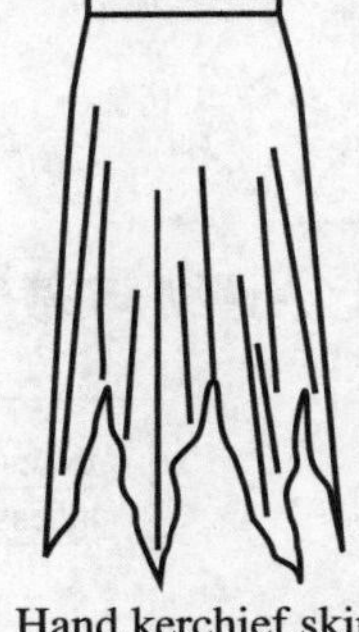
Hand kerchief skirt
手帕式裙

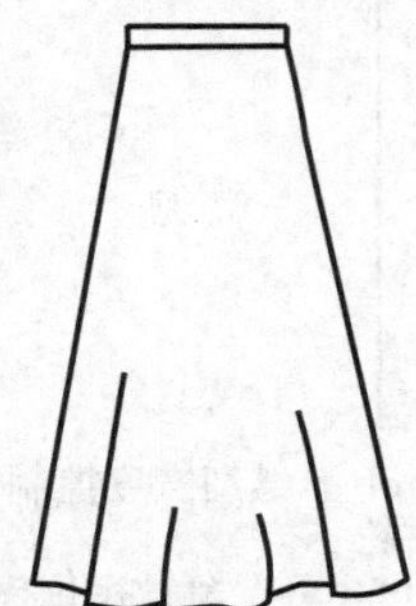
Skirt with extra flare
大摆裙

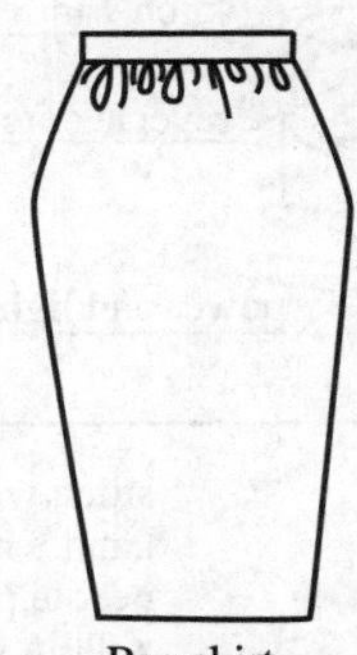
Peg skirt
螺旋裙

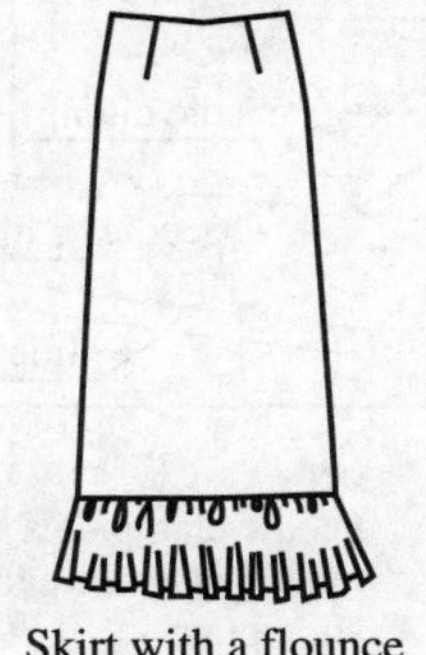
Skirt with a flounce
带荷叶边裙

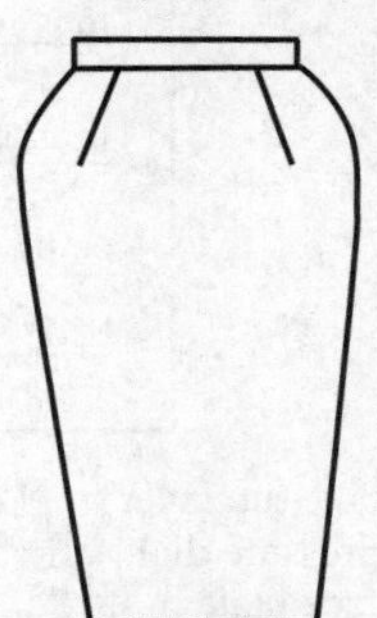
Tube skirt
筒裙

Yoke skirt
育克裙

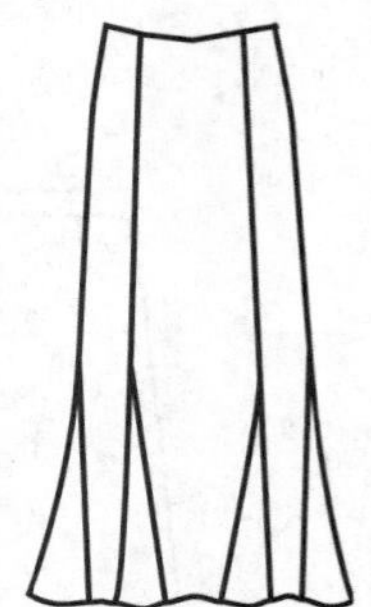

Skirt with godets
加三角布裙

六、常见帽子的种类

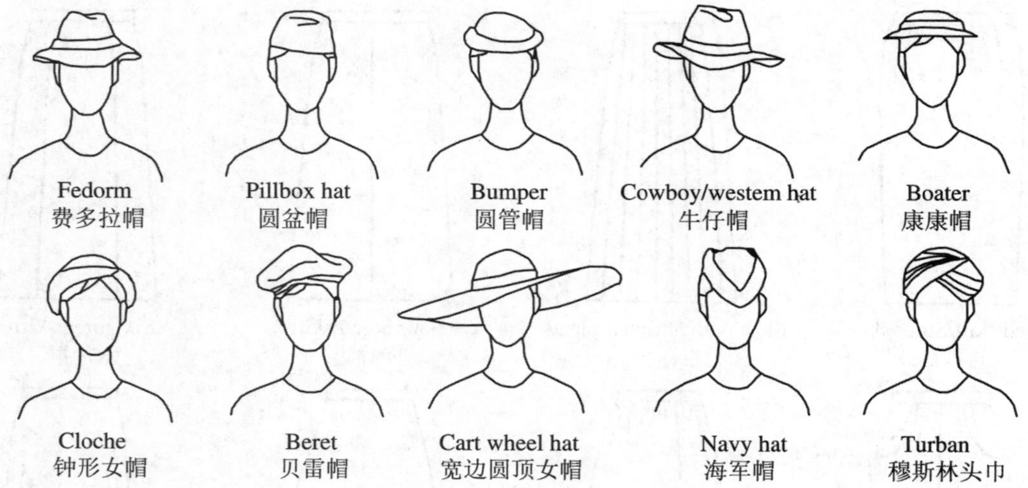

Fedorm 费多拉帽	Pillbox hat 圆盆帽	Bumper 圆管帽	Cowboy/westem hat 牛仔帽	Boater 康康帽
Cloche 钟形女帽	Beret 贝雷帽	Cart wheel hat 宽边圆顶女帽	Navy hat 海军帽	Turban 穆斯林头巾

七、缝纫与裁剪辅助设备

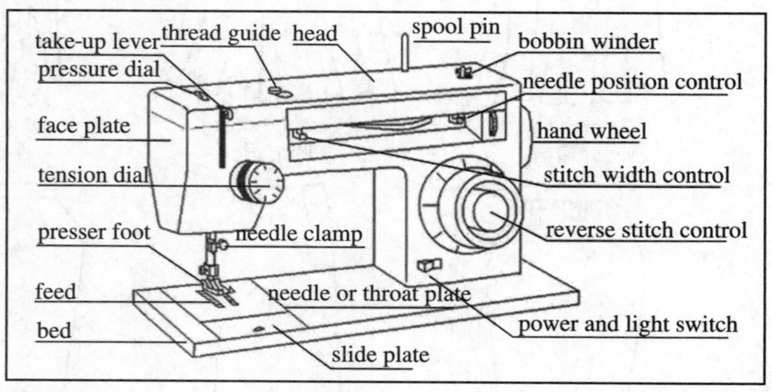

take–up lever　挑线杆
pressure dial　压辊
face plate　台板，面板
tension dial　压线板
presser foot　压脚
feed　送布
bed　底板
needle clamp　针夹
needle or throat plate　针板
slide plate　滑板
power and light switch　松紧开关
reverse stitch control　倒缝控制
stitch width control　缝纫宽度控制
hand wheel　手轮
needle position control　缝针定位控制
bobbin winder　绕线器
spool pin　线架
head　机头
thread guide　导线架

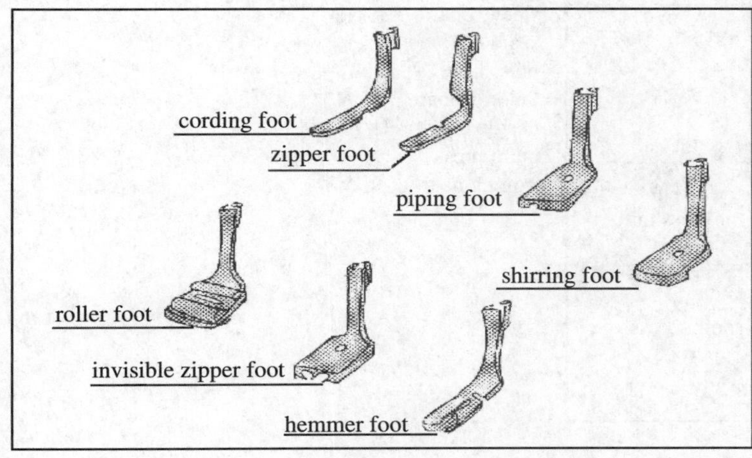

cording foot　嵌线压脚，滚边压脚
zipper foot　拉链压脚
piping foot　滚边压脚
shirring foot　褶裥压脚
roller foot　边固定轮压脚
invisible zipper foot　形拉链压脚
hemmer foot　卷边压脚

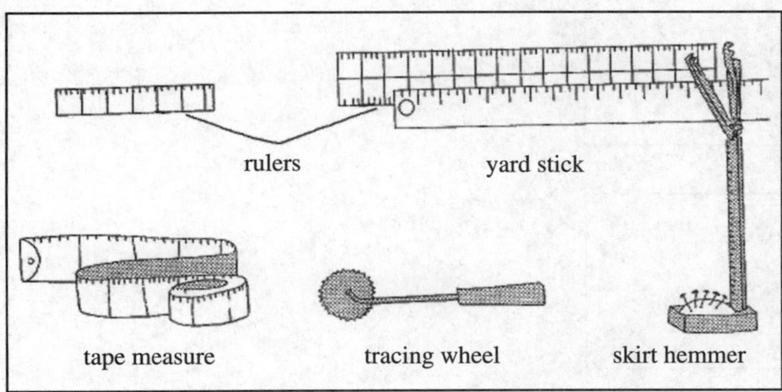

rulers　尺子
yard stick　码尺
tape measure　卷尺
tracing wheel　描线轮
skirt hemmer　裙下摆器

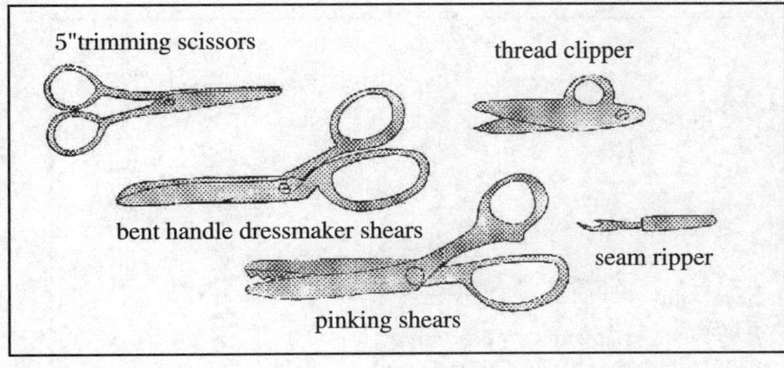

5" trimming scissors　5英寸裁缝剪刀
thread clipper　线剪
bent handle dressmaker shears　弯把裁缝剪刀
pinking shears　锯齿剪
seam ripper　拆线器

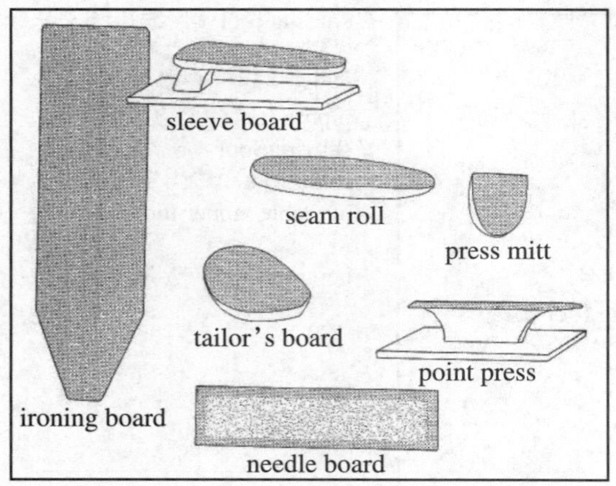

sleeve board　压袖板
seam roll　袖馒头
press mitt　手套式烫垫
tailor's board　熨衣板，马凳
needle board　针毯烫垫
point press　马凳
ironing board　熨衣板

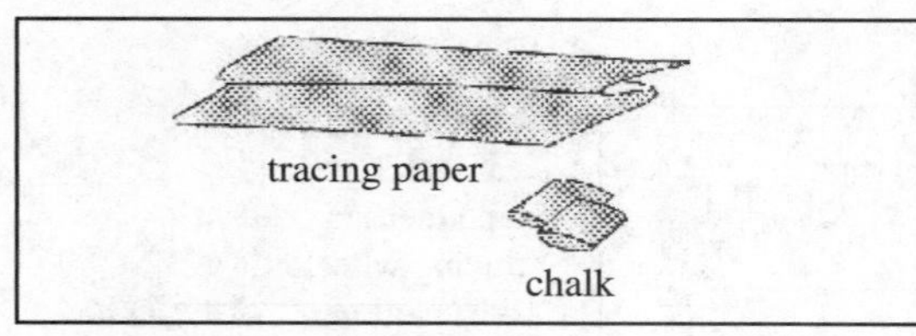

tracing paper　描图纸
chalk　划粉

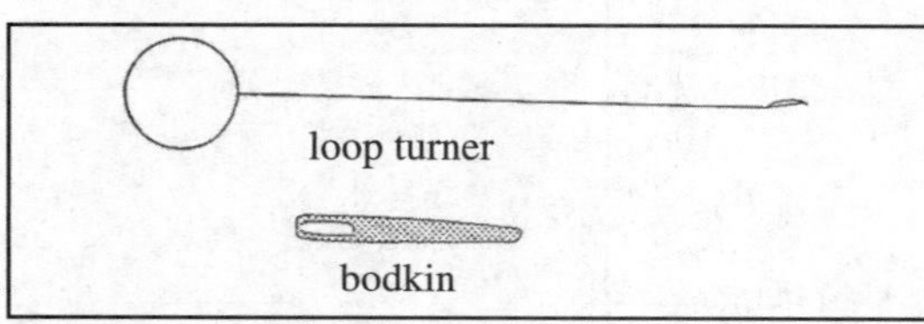

loop turner　翻带器
bodkin　锥子，钝针

八、纽扣与拉链

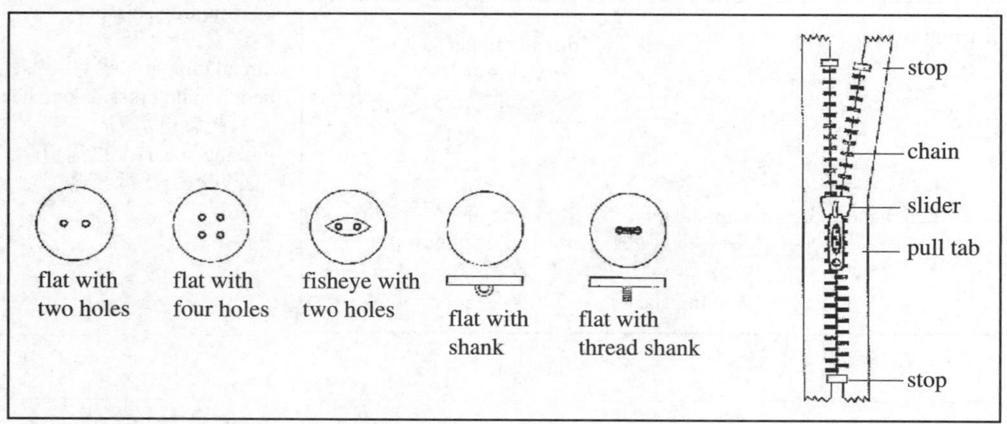

flat with two holes　双孔的平扣
flat with four holes　四孔的平扣
fisheye with two holes　双孔鱼眼扣
flat with shank　有柄的平扣
flat with thread shank　有线扣脚的平扣

stop　止端
chain　锯齿
slider　拉头
pull tab　拉片

九、国际通用的标志

PURE WOOL（纯羊毛）
（The "wool-mark" is a certification mark owned by The Wool Bureau, Incorporated.）（"羊毛标志"是国际羊毛局认证标志）

IT'S A NATURE WONDER（自然的奇迹）
（THE NATIONAL COTTON COUNCIL OF AMERICA.）（美国国家棉花总会）

十、洗涤标识

30°C
MACHINE WASH
可以机洗
HAND WASH
适宜手洗
6
30°C
MACHINE WASH
GENTLE CYCLE
可以机洗　用温和转速
7
30°C
MACHINE WASH
WOOL CYCLE
可以机洗　用羊毛衫转速
WITH SIMILAR COLORS
与同色衣物一起清洗
P
DRY CLEAN ONLY
只可干洗
P
PROFESSIONAL
DRY CLEAN ONLY
专业干洗
DO NOT BLEACH
不可漂洗
TUMBLE DRY LOW
低温干衣
TUMBLE DRY MEDIUM
中温干衣
DO NOT TWIST
OR WRING
不可扭干
DRY FLAT
平放干衣
HANG FLAT
挂干
DO NOT IRON
不可整熨
COOL IRON
低温整熨
WARM IRON
中温整熨

参 考 书 目

［1］ Maurice J. Johnson，Evelyn C. Moore. Apparel Product Development. London：Prentice Hall，1998.

［2］ G. A. Berkstresser，D.R. Buchanan. Automation in the Textile Industry from Fibers to Apparel. New York：Hyperion Books，1995

［3］ Valerie Mendes，Amy De La Haye. 20th Century Fashion. New York：Thames & Hudson. 1999.

［4］ Colin Gale & Jasbir Kaur. Fashion and Textiles. New York：Berg. 2004.

［5］ Taryn Benbow－Pfalzgraf. Contemporary Fashion. New York：St. James Press，2002.

［6］ Elizabeth Liechty，Judith Rasband，Della Pottberg－Steineckert. Fitting & Pattern Alteration. New York：Fairchild books，2010.